Introduction to
Brewing and
Fermentation
Science

Essential Knowledge for Those
Dedicated to Brewing Better Beer

Introduction to
Brewing and
Fermentation
Science

Essential Knowledge for Those
Dedicated to Brewing Better Beer

Edited by

John D Sheppard, PhD
North Carolina State University, USA

World Scientific

NEW JERSEY · LONDON · SINGAPORE · BEIJING · SHANGHAI · HONG KONG · TAIPEI · CHENNAI · TOKYO

Published by

World Scientific Publishing Co. Pte. Ltd.

5 Toh Tuck Link, Singapore 596224

USA office: 27 Warren Street, Suite 401-402, Hackensack, NJ 07601

UK office: 57 Shelton Street, Covent Garden, London WC2H 9HE

British Library Cataloguing-in-Publication Data
A catalogue record for this book is available from the British Library.

INTRODUCTION TO BREWING AND FERMENTATION SCIENCE
Essential Knowledge for Those Dedicated to Brewing Better Beer

ISBN 978-981-122-531-4 (hardcover)
ISBN 978-981-122-532-1 (ebook for institutions)
ISBN 978-981-122-533-8 (ebook for individuals)

For any available supplementary material, please visit
https://www.worldscientific.com/worldscibooks/10.1142/11966#t=suppl

Typeset by Stallion Press
Email: enquiries@stallionpress.com

Contents

Chapter 1   Introduction   1
John Sheppard

Chapter 2   Brewhouse Operations   13
John Sheppard

Chapter 3   Yeast and Fermentation   33
John Sheppard

Chapter 4   Beer Finishing   61
John Sheppard

Chapter 5   Malting   79
Sebastian Wolfrum

Chapter 6   Hops in Beer   101
Alicia Munoz Insa and Christina Schöenberger

Chapter 7   Water Quality and Usage   127
John Sheppard

Chapter 8   Beer Chemistry and Testing   149
Volker Bornemann and John Sheppard

Chapter 9   The Craft Beer Industry   173
Bart Watson

Author Biographies   199
Index   203

Chapter 1

Introduction

1.1 A Process Approach

To become truly skilled in the art and science of brewing, the brewer must develop a full appreciation of the fact that the production of beer is a biological process. The raw materials are of agricultural origin and the catalyst responsible for the almost magical transformation of simple grains into the most likable of beverages is composed of living cells. It is this inherent biological basis of the process that has attracted brewers to their art over the many centuries. It is the vagaries and sometimes confounding unpredictability of life that binds the brewer to the brewing process, in what often seems to be a never-ending challenge to harness the process to their own will.

Over the past century and a half, art has merged with science thereby revealing some of the apparent secrets behind achieving consistently high-quality beer. However, it has also revealed the amazing intricacy and complexity of each stage of the process, from malting of the barley, through production of the brewer's wort and fermentation into a delicious product to be proud of. It is this complexity that is responsible for the almost unlimited opportunity for variation in beer styles and character, but also poses great challenges to the brewer in controlling the consistency of products within the limits expected by a progressively more educated consumer.

It is the purpose of this book to develop in the reader a solid basis for appreciating and understanding the underlying scientific principles that are relevant to each step in the brewing process, encouraging systematic troubleshooting with a firm knowledge of the interrelationships between chemistry and microbiology, and how the process can be affected both positively and negatively by the actions of the brewer through adjustment of process parameters.

1.2 The Historical Perspective

It is common knowledge that brewing is an ancient art, practiced for many centuries before even the most rudimentary understanding of the underlying principles. It is a tribute to these ancient brewers that the essential details of the basic process were adapted to local conditions and transferred from generation to generation. Although most European beer produced before the 16th century could be considered as "homebrew," larger scale production occurred as early as the 8th century within the Benedictine monasteries and became more widespread throughout central Europe and England in the 13th century.

The first significant innovation was probably the use of hops as a flavoring agent and, more importantly, as a preservative. Although hops had likely been used in certain localities for several hundred years, it wasn't until the brewers in northern Germany in the 13th century fully recognized their value in prolonging beer quality that the export of German beer by sea to the Benelux countries became possible. As beer became an increasingly important economic factor in society, the next 400 years saw growth of the brewing industry throughout Europe and is considered by Unger to be the Golden Age of brewing. This expansion was eventually curtailed by competition from distillers and the increased availability of colonial beverages such as coffee, cocoa and tea.

Brewing was the first food or beverage industry to be the subject of government regulation. In 1447, the city of Munich issued an ordinance that restricted brewers to the use of only water, barley and hops, forming the basis for the famous *Reinheitsgebot*, or German Purity Decree, which was enacted in 1516 by Duke Wilhelm IV of Bavaria. After unification with Prussia in 1871, it was extended to the entire German Empire. To some extent the decree was also in protection of the bread industry, by restricting the use of rye and wheat to baking. However, there was an exception granted for a limited quantity of wheat beer to be produced. Originally, it was granted under exclusive license to the Degenberger family until they died out in 1602. After which the Duke's own family, the Wittelsbachs, ran the Munich wheat beer brewery, until leasing it to private brewers starting in 1802. Although still in force, the *Rheinheitsgebot* has been modified several times:

- its application strictly to bottom-fermented lager beers (wheat is commonly used in German ales such as Altbier and Kölsch)
- the inclusion of yeast as the fourth ingredient when the German Beer Tax law was implemented by the Weimar Republic in 1919
- the European Union (EU) ruled in 1987 that it violated free-trade and was subsequently modified to allow nonconforming beer to be imported from other EU countries

The history of brewing closely parallels important technological developments as scientists and engineers have worked together to increase the quality and consistency of beer, extend its shelf life and reduce its production costs, especially with respect to the use of manual labor. Figure 1.1 depicts a timeline between the *Rheinheitsgebot* and the modern craft beer industry in the 21st century, showing dates of some of the major events. Perhaps the single most important

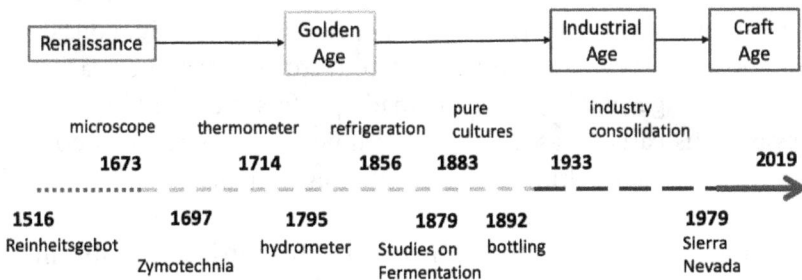

Figure 1.1. The modern historical ages of brewing and some important technological advances and events that helped to shape the industry.

development was the ability to select, store and propagate pure cultures of different species and strains of yeast. Before this became the accepted practice among brewers, the success of the fermentation process was largely left to chance.

Mixed culture fermentations were the norm as each new batch would likely contain a different mix of both yeast and bacteria. Although there was some degree of quality control based on the frequency of yeast purchasing and in-house propagation, contamination was common and highly dependent on the environment maintained within the specific brewery. The industrial brewing industry that dominated the 20th century could not have developed to the extent that it did without implementing a high level of control over the purity of their yeast cultures. During the 19th century there were many individuals who contributed to this technology, including Schwann, Pasteur, Petri, Hansen and others.

The essential issue that needed to be resolved was whether the so-called "germ theory" of fermentation was true or whether fermentation was essentially a series of chemical reactions that could occur spontaneously as a result of the reactions of the "ferment," even in the absence of living cells. The existence of microscopic living cells had been known since first observed by the Dutch scientist Antonie Leeuwenhoek in the 1670s and reported by the Royal Society in 1673 when they published his letters. However, it took another two centuries before the role of yeast cells in fermentation was definitively proven.

Perhaps the strongest advocate of the chemical explanation for fermentation was the German chemist, Justus von Liebig, considered the founder of organic chemistry. A brilliant scientist and teacher, in 1824 at the age of 21 Liebig was made professor at the University of Giessen. His career was notable for numerous important discoveries with regard to agricultural and food chemistry in particular. He developed the "law of the minimum" with respect to the growth of plants, observing that growth is always limited by the nutrient available in the least supply and promoted the use of inorganic nitrogen as a fertilizer for crops. Perhaps most relevant to the debate on the process of fermentation was von Liebig's belief that living processes could be reduced to a series of chemical reactions and that yeast cells were the product of a reaction between oxygen and organic nitrogenous matter. However, the mid-19th century saw a growing number of cell biologists who were in fundamental opposition to this purely materialistic view of life.

Theodore Schwann, born in Germany in 1810, graduated with a medical degree from the University of Berlin in 1834 before working with the well-known physiologist Johannes Peter Müller and then, in 1836, joining the Catholic University of Leuven in Belgium as a professor. While at Leuven, Schwann in collaboration with Matthias Schleiden, a German botanist, developed a "cell theory" that described the similarity between different cells and how they functioned individually as fundamental units within an organism, both plant and animal. His work also extended to the study of sporulation in yeast and added evidence for the role of yeast in the fermentation of sugar. However, 20 years later, at least in scientific circles, it remained controversial as to the mechanism of fermentation and whether living cells were indeed responsible. Contemporaries on opposite sides of the issue were two Frenchmen, Louis Pasteur and Pierre Eugene Berthelot.

Louis Pasteur, born in 1822 in Dole, France, began his academic career in 1848 at the University of Strasbourg as a professor of chemistry, focusing his efforts on the optical properties of tartaric acid and eventually discovering the concept of chirality. However, it was not until 1854 when he became the Dean of the Faculty of Sciences at the University of Lille that he began his work on fermentation. Three years later, in 1857, Pasteur became director of scientific studies at the *École Normale* in Paris and in that same year published his first paper on fermentation. Even at this early stage, based on his experimental evidence, his conclusion was profound — in addition to the production of alcohol and carbonic acid, a portion of the sugar that was fermented in the presence of yeast was used by the yeast to increase the mass of cells present in the ferment. He argued that if the fermentation was simply a spontaneous reaction, then none of the material would be incorporated into the total mass of yeast. In a second paper, published in 1860, he confirmed this hypothesis with additional data, stating categorically that fermentation was a physiological process, but also stating that he had no understanding of the underlying chemical mechanisms.

One of the main proponents of von Liebig's chemical theory of fermentation and putrefaction was Pierre Berthelot. Like Pasteur, Berthelot was also from France, the son of a physician, in 1859 he attained the position of Professor of Organic Chemistry at the *École Supérieure de Pharmacie*. His work focused on synthetic organic chemistry and the role of enzymes in biological systems. He maintained that the "ferment" that is responsible for the inversion of cane sugar (sucrose) into fruit sugar (fructose) is produced by yeast but is active even in the absence of living cells, i.e., from an extract of yeast. As both Berthelot and Pasteur began publishing their work on fermentation beginning in the late 1850s, the difficulties of defining what composed the "ferment" and distinguishing between enzymatic action and the metabolism of whole cells were evident.

Without a clear understanding of the biochemical processes underlying both alcoholic fermentation and the growth of cells, the tug-of-war between chemists and biologists continued for more than 20 years. As the evidence increased in favor of the biologists, the issue was largely settled by 1876, coinciding with the publication of Pasteur's landmark book *Études sur la Biére*, in which Pasteur describes many of his previous experiments and observations on the relationship between yeast and fermentation.

Although the book presents a wealth of information collected from years of meticulous experimentation, most notable were Pasteur's observations on:

- beer spoilage from contamination by unwanted microbes and the destruction of these organisms by boiling the wort
- spontaneous wine fermentation — proving it will not occur if the grape is boiled and only inoculated with an extract from inside the grape, disproving Liebig's theory on the spontaneous generation of living cells by oxidation of "albuminous" substances
- different yeasts responsible for alcoholic fermentation, specifically top fermenting ale yeast (referred to as "high" yeast) in

comparison to the bottom fermenting lager yeast ("low" yeast) and methods to maintain culture purity

- the effects of oxygen on yeast and the differences between aerobic and anaerobic metabolism
- a new process for the production of beer — emphasizing the importance of controlling oxygen concentration in the wort, preventing microbial contamination and using a pure culture of yeast for fermentation

The concept of using pure cultures for brewing required the means for culturing and storing microbes in a laboratory environment in a manner that would both maintain purity and reveal if a contamination had occurred. This was greatly facilitated by the development of the Petri plate (sometimes referred to as a *dish*) and the use of a semisolid agar-based substrate to support microbial colony formation, providing the added benefit of allowing for quantification of the number of viable cells in a measured volume of sample (Figure 1.2).

Julius Petri, born in Barmen Germany in 1852, received a medical degree in 1876 before becoming an assistant to Robert Koch, the physician and scientist well-known for his study of contagious diseases

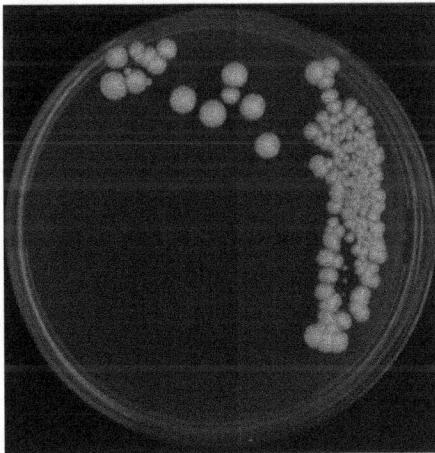

Figure 1.2. A Petri dish showing individual yeast colonies growing on nutrient agar.

and the development of vaccines. While in Koch's laboratory, he perfected the technique of culturing bacteria on agar with the use of his special shallow round glass culture dishes, referred to as Petri plates. Although now largely replaced with a disposable plastic variety, use of Petri plates for performing "plate counts" remains the gold standard for enumerating viable cells in a sample, under the assumption that each viable cell is responsible for forming a single visible colony on the surface of the agar. Thus, plate counts are stated in terms of colony-forming units or CFUs. This development coincided with Hansen's work in isolating pure species of yeast in the Carlsberg Laboratory in Copenhagen.

Emil Christian Hansen, born in 1842, was a Danish mycologist (fungi specialist) who was hired by the Carlsberg Laboratory in 1879 and in 1883, claimed to have been the first to isolate in pure form the species of yeast now used for lager beer production, *Saccharomyces carlsbergensis*. It is now known that this was the same species of "low yeast" Pasteur had referred to years earlier as *Saccharomyces pastorianus*, the name that is in use today. The Carlsberg Laboratory was in the business of producing and selling yeast, so in the late 19th century, pure cultures of lager yeast were soon available to brewers throughout Europe and North America.

Thus, the culmination of almost a half century of progress by microbiologists led to the adoption of new brewing practices, which in turn set the ground work for a new age of brewing in the 20th century — characterized by improvements in both quality and consistency, beer became a high-volume commodity with many markets dominated by a few multinational companies with global distribution. It wasn't until the late part of the century that small breweries once again became established, especially in the USA. This craft-beer movement continued to expand into the 21st century and is analyzed in detail from a business perspective in Chapter 9.

1.3 The Brewing Process

A process is typically defined as a transformation of one or more substances, or raw materials, into one or more other substances, referred to as products or byproducts. The transformation also implies that energy is involved and therefore both energy inputs and outputs occur. Processes are often categorized as either *batch* or *continuous*, based on whether the process occurs over a finite time period, or continues indefinitely. Most processes consist of multiple steps that are more or less discrete and can be analyzed separately from other steps in the overall process. However, caution must be exercised in this approach as the outputs from one step will be the inputs to the succeeding step. Engineers often refer to process steps as *unit operations* that can be used in any number of different types of processes. To some extent this is also true for the brewing process, as unit operations such as milling and fermentation are often used elsewhere. This book describes each of the unit operations individually, however, the reader must appreciate that brewing is a holistic process in which the performance of each operation has consequences for the performance of most of the subsequent operations. Although there are distinct advantages to continuous processes, the vast majority of brewing is carried out in individual batches and the unit operations will be described in that context.

Unit Operations

Historically, especially in Europe during the 19th century and before, most breweries had a high degree of vertical integration. To ensure a reliable supply of raw materials, the brewery would either own the land for barley production or would have long-term contracts with local farmers to supply their barley needs. This required the brewery to also operate a malt house, producing the specific types of malt required by their own beer recipes. However, since the advent of large malting companies, breweries generally no longer find it economical to do their own malting so various types of malt are purchased on an as-needed basis. Even so, an understanding of malting technology and its impact on malt characteristics is still important for a brewer. It helps in designing recipes and selecting the appropriate varieties to obtain specific effects in the beer. The process of malting is described in detail in Chapter 5.

Although there are many differences in detail for each unit operation involved in the brewing process, the standard order of operations as shown in Figure 1.3 and described in Table 1.1, are milling, mashing, lautering, boiling, fermentation, maturation, clarification, carbonation and packaging. The first four operations are carried out to

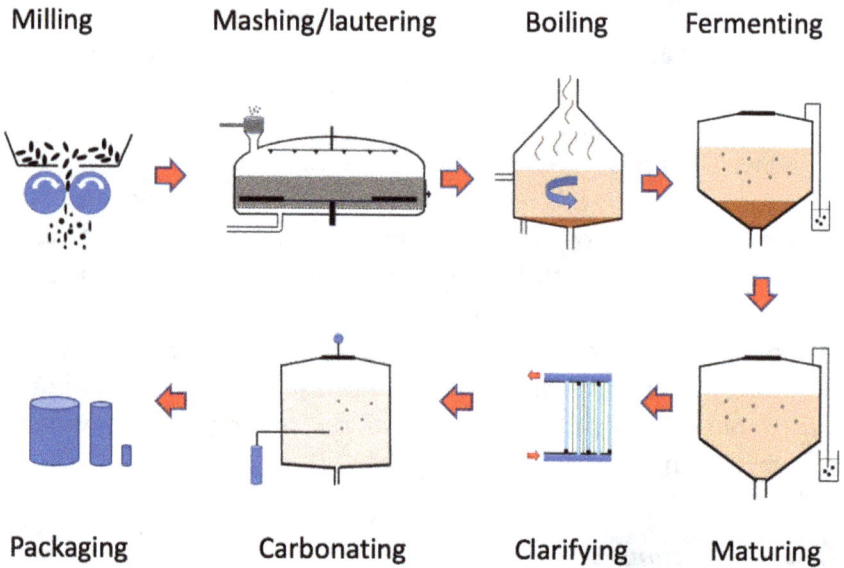

Milling   **Mashing/lautering**   **Boiling**   **Fermenting**

Packaging   **Carbonating**   **Clarifying**   **Maturing**

Figure 1.3. The principal unit operations involved in the brewing process. Mashing/lautering may be carried out in separate vessels, while the kettle is often used to both boil and whirlpool the wort.

Table 1.1. The major material inputs and outputs for the various unit operations comprising the brewing process.

Unit operation	Material inputs	Material outputs
1. Milling	Dried malt	Ground malt
2. Mashing/lautering	Ground malt	Sweet wort, spent grain
3. Boiling	Sweet wort, hops	Sweet wort, trub
4. Fermentation	Wort, yeast	Beer, CO_2, yeast slurry
5. Maturation	Beer (with yeast)	Beer, CO_2, yeast
6. Clarification	Beer (with yeast)	Clear beer, yeast paste
7. Carbonation	Clear beer, CO_2	Finished beer

transform the malt into a fermentable liquid form, referred to as *wort*, and are described in detail in Chapter 2, Brewhouse Operations.

The other basic ingredients in addition to malt are water, hops and yeast. Yeast is the biological catalyst responsible for fermenting the wort into the ethanol, CO_2, glycerol and many other trace compounds that affect the character of the beer. Yeast metabolism in relation to fermentation is described in Chapter 3. As one of the principal ingredients, hops are the traditional bittering agents added during the boiling of the wort and their characteristics and use are described separately in Chapter 6, Hops in Beer. The operations that are performed post-fermentation, maturation, clarification, carbonation and packaging are included in Chapter 4, Beer Finishing. Water chemistry can have a large impact on the various process steps and also the character and quality of the beer product. Water composition, treatment and sanitation is covered in detail in Chapter 7, Water Quality and Usage. Related to water chemistry, the underlying chemistry of beer as related to its testing and quality control is described in Chapter 8, Beer Chemistry and Testing.

1.4 Topics for Discussion

- How do biological processes differ from chemical processes?
- Does the *Reinheitsgebot* still have relevance to the modern brewing industry in Germany and elsewhere?
- Why did it take so long to prove the germ theory of fermentation?
- Why has the use of pure yeast cultures in wine production not been as widely adopted as in brewing?
- Why has the use of continuous processing, in contrast to batch processing, rarely been used for beer production?
- How would you rank the various unit operations as to their relative impact on beer quality?

1.5 Further Reading

1.   Barnett JA. A history of research on yeasts 2: Louis Pasteur and his contemporaries, 1850–1880. *Yeast*. 2000;16:755–771. doi:10.1002/1097-0061 (20000615)16:8<755

2. Encyclopaedia Britannica Editors 2019. *Theodore Schwann, German Physiologist.* www.britannica.com/biography/Theodor-Schwann

3. Holle SR, Schaumberger M. *The Reinheitsgebot — One Country's Interpretation of Quality Beer.* MoreFlavor Inc.; 2019. www.morebeer.com

4. Pasteur L. *Studies on Fermentation.* Reprint Edition. Cleveland, OH: BeerBooks.com; 2005.

5. Petri JR. 2019. https://en.wikipedia.org/wiki/Julius_Richard_Petri

6. Unger RW. *Beer in the Middle Ages and the Renaissance.* Philadelphia PA: University of Pennsylvania Press; 2004.

7. von Liebig J. 2019. https://en.wikipedia.org/wiki/Justus_von_Liebig

https://doi.org/10.1142/9789811225321_0002

Chapter 2

Brewhouse Operations

2.1 Malt Grinding (milling)

The malt is almost always delivered to the brewery in bulk (25 kg bags or by truck or train and transferred into a hopper) as intact, uncracked grains. In the uncracked state the dried malt is resistant to moisture and odors, thus retaining its quality during storage. However, efficient mashing of the malt (see Section 2.2) requires that the grains be milled in a way that allows for full exposure of the starch, proteins and enzymes to water. In theory, this is best accomplished with a fine grind to maximize surface area of the starchy endosperm. But the grind cannot be too fine, otherwise the husk fragments will be too small and porosity will be inadequate to act as an effective filter bed during lautering. Thus, the milling operation needs to be optimized to give specific ratios of grain fractions. For example, with a two-roller dry mill as illustrated in Figure 2.1, a typical distribution might be 50% fine grits, 30% coarse grits, 10% flour and 10% husk (see Table 2.2). If a mash filter is used instead of a lauter tun, then a finer grind, as produced by a hammer mill, is advantageous for improving extraction efficiency.

The performance of a dry milling operation can be evaluated with the use of a series of sieves or screens with differing mesh sizes. In this way the ground malt is fractionated into husks, coarse grits, fine grits and flour, and each fraction can be weighed to determine the relative

Intact dried malt grains

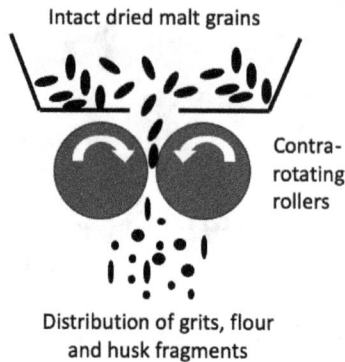

Contra-
rotating
rollers

Distribution of grits, flour
and husk fragments

Figure 2.1. A roller mill cracks open the husks and produces a distribution
of fragments of different sizes.

percentage produced during milling. The Standard Method as published by the American Society of Brewing Chemists (ASBC) specifies
the use of the following sieves (Table 2.1):

Table **2.1.** ASBC standard sieve sizes for
determining the performance of dry milling.

Sieve number	Mesh width (mm)	Milled fraction
10	2.000	Husk
14	1.410	Husk
18	1.000	Husk
30	0.590	Coarse grits
60	0.250	Fine grits
100	0.149	Flour
Pass through	—	Fine flour

Roller mills are of different designs, depending on the number of
rollers and whether intermediate sieving is employed. The distribution of fractions after milling can be altered based on the number of
rollers, the diameter of the rollers, adjustment of the spacing between
rollers and their speed of rotation. Optimal settings for the milling
may need to be readjusted for different types of grains (malted barley

versus malted wheat), the degree of modification of the malt and also for specific mash or lauter tuns that may utilize different height/diameter ratios.

One of the inherent disadvantages to a dry milling process is the significant volume of dust that is produced. This flour dust can be hazardous to breath and can pose a potential explosion risk. The mill is therefore usually enclosed in a separate area within the brewery that is ventilated to the exterior of the building and the milled grain is transported by auger to an intermediate storage hopper prior to mashing.

An alternative to dry milling and often used in larger breweries, is wet milling. This involves adding a conditioning step prior to the milling to partially hydrate the kernels. The conditioning of the grain is accomplished by wetting the kernels, either by a short 1- to 2-minute exposure to water or steam or by a longer steeping for up to 30 minutes. These two conditioning processes are carried out quite differently. The duration of the short wetting period is controlled by the residence time of the grain in a screw-type conveyor, which transports the grain to the grinding mill. In this case, the moisture content increases by only 1–2%, compared to up to 30% during the 30-minute steeping, which takes place in a recirculated vat. The limited degree of hydration during the short conditioning results in the husks becoming softer while the endosperm remains dry. In the subsequent wet milling process, the husks remain largely intact, while the crushed endosperm is fully exposed in the mash. During steep conditioning the kernels become hydrated to about 30% moisture content, resulting in an endosperm with a paste-like consistency. As the moistened kernels are passed through the roller mill, the endosperm is squeezed out of the husk, which has been toughened during steeping. As with the short conditioning, the husk remains mostly intact, improving the run-off rate during lautering. There is also the potential for improving the extract efficiency, especially if the steep water is used in the mash.

Another variation on the milling process, referred to as spray steep roller milling, involves the integration of conditioning, milling and mashing-in all in one continuous process. The conditioning of the dry kernels is very similar to the standard wetting process with exposure times to a spray of hot water limited to between 1 and 2 minutes.

After conditioning, the kernels are immediately ground in either a 2-row or a 4-row mill and fed directly into the mashing vessel.

2.2 Mashing

The substrates present in the ground malt must be enzymatically transformed and dissolved into a liquid form, referred to as the sweet wort, which can then be fermented by the yeast. This process is accomplished by subjecting the malt fractions to a precise amount of water at controlled temperatures to promote the enzymatic reactions for a specific period of time. The objective is to maximize the yield of fermentable substrates, while providing a liquid wort with characteristics that will ultimately form the basis for the particular style of beer being brewed. Beer characteristics such as flavor, aroma, color, body and ethanol concentration are all affected by the mashing process. The principal variables with regards to an all-grain mash are:

1. Malt type(s) and degree of modification
2. Mash temperature-time profile
3. Mash thickness as reflected by the grist/water ratio
4. Mash water composition, especially the mineral content and pH

The use of sources of fermentable sugar other than malt, referred to as adjuncts, is a widespread practice and has consequences for how the mashing is performed. This topic will be dealt with later in the chapter. Chapter 5 describes what is involved in the malting process and how it affects the performance during mashing.

Enzymatic Reactions

The important groups of enzymes that are activated during mashing are the amylases, proteases and phytases. These enzymes are responsible for starch degradation into fermentable sugar, production of free-amino nitrogen (FAN) from malt proteins and the release of inorganic phosphate, which results in a decrease in the mash pH. In addition to these, the enzyme β-glucanase hydrolyses β-glucans, polymeric

Table 2.2. Important mash enzymes and the effects of temperature on activity.

Enzyme	Optimal temp range °F	Optimal temp range °C	Max. efficiency °F	Max. efficiency °C	Denatures at °F	Denatures at °C
Phytase	86–128	30–53	95	35	~140	60
β-glucanase	95–131	35–55	113	45	~140	60
Peptidase	113–122	45–50	122	50	~145	63
Proteinase	122–140	50–60	136	58	~155	68
β-amylase	126–144	52–62	144	62	~160	71
α-amylase	149–153	65–67	153	67	~170	77

materials that originates in the cell wall of the barley and this can contribute to an increase in the wort viscosity to the point that run-off during lautering and/or filtration may be hampered. Table 2.2 lists the major enzymes and the effects of temperature on activity.

The starch in the malt is present in the following two forms:

- Amylose, a linear chain of between 200–1000 glucose molecules with α 1–4 linkages, accounts for about 25% of the total starch. It has one reducing end and one non-reducing end (OH group) per molecule.
- Amylopectin, a branched chain of between 40 and 70 amylose segments of 25 glucose molecules each, joined by α 1–6 linkages. It has one reducing end and many nonreducing ends. It is tightly packed and mostly inaccessible to enzymes until gelatinization occurs.

The enzyme β-amylase hydrolyses starch from the nonreducing end into molecules of the disaccharide maltose, which is composed of two molecules of glucose. It is referred to as the "saccharifying" enzyme and is the primary determinant of the fermentability of the wort. It has optimal activity at a pH of 5.7 and between temperatures of 55 and 62°C, although it retains some activity up to 71–72°C. It is able to attack alternate 1–4 linkages once the starch has become gelatinized at about 60°C, however, it cannot get past the 1–6 linkages. So, based on β-amylase alone, this would result in almost complete conversion of the amylose into maltose while amylopectin would be only partially hydrolyzed into maltose, leaving the majority as larger molecular weight dextrins, which are not fermentable. Fortunately, another class of enzyme, α-amylase, is able to bypass the 1–6 branch points and attack internal 1–4 linkages, thereby "liquifying" the gelatinized starch and providing ready access for β-amylase enzymes. Although the α-amylase works much slower, it is the primary determinant of the final extract yield and the amount of residual dextrins in the wort, typically 20–40% of the initial starch content. The ideal temperature is slightly higher than that for β-amylase, 65–67°C, and the ideal pH is about 5.3.

In addition to the wort having the correct amount of fermentable sugar, success of the fermentation relies on the yeast having access to sufficient nitrogen. Most of the accessible nitrogen will be in the form of amino acids (FAN) with a small amount of ammonia. The malting process, as described in Chapter 5, results in significant protein modification and the production of soluble proteins. These soluble proteins are further degraded during mashing by the action of the protease enzymes naturally present in the mash. There are two types of proteases in action: the proteinases, which degrade intact proteins into peptides, and the peptidases, which further convert these into smaller peptides and amino acids that are readily available to the yeast. Both types of protease enzymes have an optimum temperature range below that of the amylases, proteinases between 50 and 60°C and peptidases between 45 and 50°C. Because the protease enzymes are not very tolerant to high temperature, as part of the mashing protocol, a so-called "protein rest" may be performed at 50°C for 20 minutes to maximize the conversion of the soluble protein fraction. Although the total quantity of FAN produced is important, not all amino acids have the same impact on yeast metabolism. This topic is covered in more detail in Chapter 3.

The other important class of enzymes involved in mashing are the phytases and to a lesser extent the nucleases. These enzymes are of critical importance for establishing the pH of the mash and subsequent quality of the beer. The action of phytase results in the release of phosphate ions from the organic phosphate found in the malt. The phosphate ions combine with potassium to form two salts: dibasic potassium phosphate (K_2HPO_4) and monobasic potassium phosphate (KH_2PO_4). The pH of the mash will be largely determined by the relative quantities of these two salts. With sufficient calcium present (for example, as calcium sulfate), the alkaline potassium salt will be converted to the monobasic form, by the following equation:

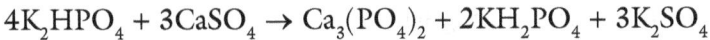

$$4K_2HPO_4 + 3CaSO_4 \rightarrow Ca_3(PO_4)_2 + 2KH_2PO_4 + 3K_2SO_4$$

The calcium phosphate precipitates from the solution resulting in rapid conversion of the dibasic salt to the monobasic form with an associated decrease in the mash pH from about 6 to between 5.2 and 5.7, a level of acidity much more conducive to good enzyme activity. However, the pH of the mash is not only affected by the action of the phytase enzymes but also by the level of calcium, magnesium and bicarbonate ions, so overall water composition is important to consider. This topic is returned to in Chapter 7.

Mashing Protocols

The different methods and equipment employed for mashing, all share the same goal: the efficient production of a readily fermentable solution of sugars and other nutrients from various substrates (principally malted grain) with a high yield. This goal can be attained using multiple strategies, each requiring somewhat different technology and procedures. Differences can even be found between individual breweries using the same equipment, where procedures have been adapted to the specific recipes or beer styles. Brewhouses are usually described based on the number of vessels employed, *i.e.,* 2-, 3- or more rarely 4-vessel systems. The number of vessels in the brewhouse defines the role(s) of each vessel and the optimal design of each vessel to achieve the desired performance. Within the brewhouse operations, the mashing protocol can be carried out in either one or two vessels, depending on whether the protocol is based on an *infusion* or *decoction* method or, whether in addition to malted grain, other adjuncts are used that require a separate vessel for cooking. Both the infusion and decoction methods share some common features:

1. A mash-in (or dough-in) step during which the grain is wetted at a specific temperature in a manner to prevent clumping and to begin activation of the native enzymes.
2. Control of the time–temperature profile in order to maximize enzymatic activity and yield of sugars.
3. A mash-out step during which the temperature is increased sufficiently to stop enzymatic activity.
4. A sparge and lautering step to recover and separate the sweet wort from the grain residue (spent grains).

Of these four common features, perhaps the largest variation between protocols occurs in the time–temperature profile and the means used to control it. The following description focusses on four principal protocols: single infusion, step infusion, decoction and double mashing with adjuncts.

The single infusion protocol as depicted in Figure 2.3, is based on a single vessel mash tun (or tub), historically used in the United Kingdom for production of English ales where an all-grain recipe is used with well-modified malts. The design of a typical mash tun is shown in Figure 2.2.

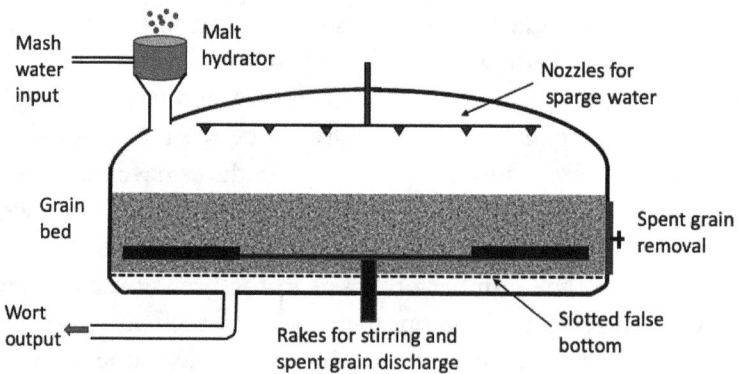

Figure 2.2.   The simplified construction of a combination mash tun/lauter tun used for a single infusion mashing protocol.

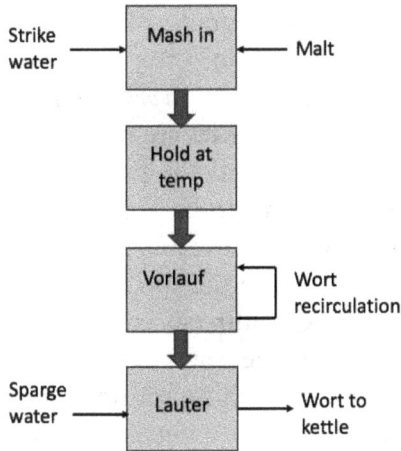

Figure 2.3.   Steps in a single-infusion mash protocol.

Note that when the mash tun is increased in capacity, the depth of the grain bed remains quite constant while the diameter is increased. This is because if the grain bed becomes too thick it will compact at the bottom and prevent adequate percolation of the liquid wort through the bed and false bottom. It is a dual-purpose vessel since in addition to providing an environment to promote the enzymatic reactions, the slotted false bottom also allows for separation of the sweet wort from the spent grains. In this protocol the mash tun is not heated and so the temperature of the *strike water* during mashing-in must be externally adjusted to achieve the desired fixed temperature in the tun. A device referred to as a *Steel's masher* is used at the inlet to the tun to effectively hydrate the grain with the strike water.

The vessel must also be well insulated to maintain the appropriate temperature for the desired mash time, typically 45–60 minutes. Since the mash temperature is fixed but optimal temperatures for each type of enzyme differ, a compromise must be made and an intermediate temperature between 64 and 70°C is typically chosen (referred to as the "brewer's window"). Once sufficient time has passed to achieve the desired enzymatic results, the sweet wort is drawn off the bottom of the tun through the slotted false bottom. Initially, this *run-off* will be cloudy from the suspended grain solids that passed through the slots and so it is pumped back into the top of the tun through the sparge arms, a process referred to as *vorlaufing* (loosely translated as the *first running*). The fine particles are mostly captured as they filter down through the grain bed and the process of vorlaufing is completed once the run-off appears to have clarified. At this stage lautering begins as the sweet wort is diverted into the kettle. However, as the liquid level falls, but before the top of the grain bed is exposed, sparge water at about 78°C is introduced through the overhead nozzles. The purpose of the sparge water is to both maintain a constant liquid head, or level, in the mash tun during lautering and also wash off the soluble sugars from the spent grain. The sparge step does, however, dilute the wort, so both the flow rate and the volume of sparge water must be accurately controlled to achieve the desired sugar concentration in the kettle prior to boiling.

A step infusion mashing protocol may also be performed in single vessels if there is an ability to either heat the vessel directly, usually with jacketed steam, or if additional volumes of hot water (*underlet* water) are added to the mash tun at appropriate intervals. In both cases, to achieve temperature uniformity within the vessel, the mash tun is usually equipped with a mixer or adjustable rakes. In this case, lautering is often accomplished in a separate vessel, the lauter tun. The advantage to the step infusion process is that the temperature of the mash can be controlled to pass through a specific profile that may or may not include constant temperature holds, or rest periods, along the way. A typical step infusion time–temperature profile is shown in Figure 2.4.

As evident from the time–temperature profile, the advantage of the step infusion protocol is that the hold periods are designed to be at temperatures that are near optimal for all major enzyme groups. This should result in better control over the enzymatic reactions and higher yields of both fermentable sugar and FAN. Also, the time and temperature for mash-out can be strictly controlled to stop enzyme activity prior to beginning the vorlaufing and sparge. The increased complexity and expense of the equipment and control systems required to implement this protocol may not be justified, especially

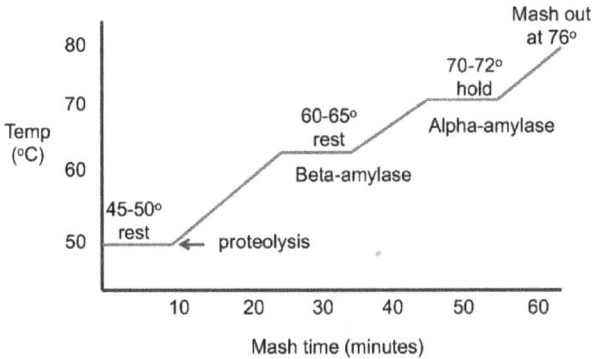

Figure 2.4.   A step infusion mash protocol to optimize enzyme activity.

for English ale varieties in which considerable residual dextrins may be tolerated. However, with the use of less well-modified malts or for the production of dry light lagers, this protocol should provide a wort with superior characteristics.

Similar results can be attained using the decoction protocol, traditionally used in Germany and Eastern Europe. As with the step infusion method, multiple rest periods at specific temperatures are employed to maximize enzyme efficiencies. The decoction method requires at least two agitated vessels in addition to a lauter tun. After mashing in at a relatively low temperature, a portion of the mash is pumped to the decoction vessel where it is heated to boiling before being returned to the mash mixer. The amount of mash that has been removed and the rate of its return, determine the temperature profile in the mash mixer for that decoction. Single or multiple decoctions are possible with this method depending on the number of rest periods at specific temperatures that are desired. Generally, there are at least two rest periods, one for proteolysis and one for saccharification, with the total mash time extending to several hours. A triple-decoction mash profile, which also includes an acid rest, is illustrated in Figure 2.5.

Once the mash is completed, the full volume is transferred to a lauter tun for separation of the spent grain from the sweet wort, including vorlaufing and sparging steps as used with other mash

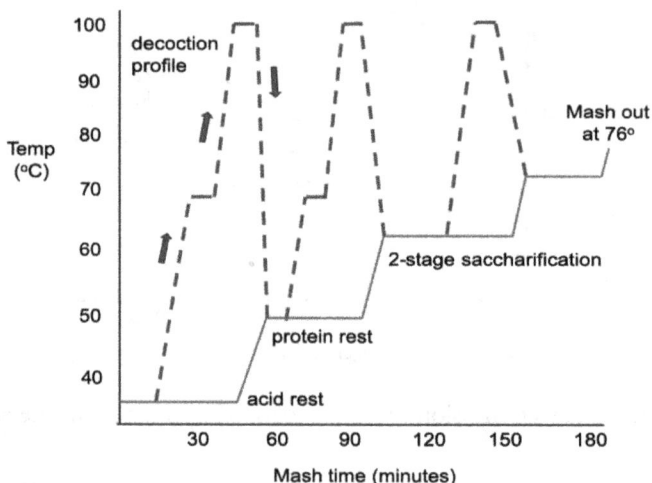

Figure 2.5.   A triple-decoction profile showing mash temperature (solid line) and decoction temperature (dashed line).

protocols. Implementing a decoction mashing protocol is relatively complex and expensive but does allow for the use of poorly modified malts and provides for a high proportion of fermentable sugars at moderate concentrations, ideal for European lagers. Although the results are very similar to that obtained with modern temperature control technology, now being employed for step infusion mashing, there can be differences in both flavor and color in the final product. So, the decoction method continues to be used by older brewing companies that wish to continue with their traditional practices, convinced that it is the best way to maintain the special characteristics of their beer.

Double mashing refers to the use of adjuncts to supplement barley malt. This has become increasingly common, especially by large national and multinational brewers. Depending on the raw material, typically either corn, rice or other unmalted cereal grains, adjuncts supply a more economical source of fermentable sugars than the malted grain and also impact the beer flavor by reducing residual sweetness. This is especially convenient for the production of high-gravity beer, in which it may not be feasible to use such a high grist/water ratio that would be necessary using conventional mashing. Solid adjuncts are mashed under quite different process conditions than malt and require a separate vessel, referred to as a cereal cooker, as the higher gelatinization temperatures would denature the malt enzymes. The conversion of the adjunct starch is also a result of enzymatic action, usually based on the addition of commercial preparations of microbial-based amylases. The time–temperature profile used in the cereal cooker will be based on the recommendations of the enzyme supplier for the type of solid adjunct being processed and the particular mixture of enzymes to be used. Once optimal conversion of the starch has been accomplished, the adjunct wort is transferred to the mashing vessel where it is mixed with the malt at the appropriate time to raise the mash temperature. Liquid adjuncts, such as cane or corn syrup, do not need cooking and can be added directly to the kettle and mixed with the wort after lautering.

Depending on the style of beer, adjuncts may be used to provide up to 40% of the fermentable sugars. But because adjuncts are lower in nitrogen than barley, yeast metabolism may be affected and the wort may require supplementation with amino acids and other nutrients.

The efficiency of the mashing operation is based on the yield of soluble sugars, both fermentable and nonfermentable, in comparison to the maximum value that could be obtained based on the starch content of the specific raw materials. Maximum yield values range between 63% for rye wheat to over 80% for rice and flaked corn, with different varieties of barley malt averaging between 75 and 80%. The actual yield obtained by the mashing process will be somewhat lower than the maximum, as affected by the degree of modification (for malt), whether additional enzymes are used and the mashing protocol. So, with overall mashing efficiencies between 85 and 95% of maximum, actual yield values for sugar from barley malt will vary between about 65 and 75%.

Knowing the actual yield for this process allows the brewer to calculate the expected original gravity (OG) of the wort based on the overall grist/water ratio, which includes sparge water. For example, if the final expected yield is 70%, then 100 kg of malt will provide

70 kg of sugar. With an overall grist/water ratio of 0.2 by weight, this would result in a sugar concentration of:

$$(0.7 \text{ kg sugar/kg malt}) \times (0.2 \text{ kg malt/kg water}) =$$
$$0.14 \text{ kg sugar/kg water}$$

Converting this to brewing units, the pre-boil specific gravity of the wort would be 1.056 or 14°Plato. Note that this pre-boil value is after sparging, so the actual grist/water ratio used during mashing will need to be higher than the overall ratio by a factor proportional to the ratio of sparge water to mash volume:

$$\text{mash grist/water ratio} = \text{overall ratio} \times (1 + \text{sparge volume/} \\ \text{mash volume})$$

Lautering

As described in the previous section on mashing protocols, lautering is the process of filling up the boil kettle at the conclusion of mashing by separating the sweet wort from the spent grains. It is performed either in a combination mash–lauter tun (Figure 2.2) or in a separate vessel. In both the cases the vessel requires a perforated false bottom that retains the solids while allowing for drainage of the liquid wort. Most lauter tuns include a mechanical raking system that gently turns over the grain bed to aid in drainage and also for spent grain removal. The design of the lauter tun also requires one or more nozzles for spraying hot sparge water onto the top of the grain bed to wash off residual sugars that adhere to the grain husks. The vorlaufing proce-dure of recirculating the first run-off through the sparging system, utilizes the grain bed as a filter to remove the fine particulates that have passed through the perforations or slots in the false bottom.

An important consideration in lautering is the rate at which the wort drains through the perforated bottom, as a very slow run-off will have a significant impact on brewhouse productivity. The major determinants of the run-off rate are the differential pressure across the grain bed, the bed permeability and the wort viscosity. The differential pressure is largely determined by the liquid height in the lauter tun, ideally maintained about 1 cm above the top of the grain bed during lautering by controlling the flow of sparge water and back pressure on the outlet. Too high a differential pressure will quickly result in bed compaction and low bed permeability leading to a slow run-off. Bed permeability will also be affected by the initial milling process and distribution of particle sizes, as well as the mashing process itself, which results in a variety of materials that can impede flow — including amorphous proteins such as glucans. Wort viscosity is a function of the concentration of dissolved substances in the wort, principally sugars, both fermentable and higher molecular weight dextrins. So, higher grist/water ratios during the mash will increase viscosity and slow the run-off, as will poor conversion efficiency. Because viscosity is inversely correlated with temperature, the use of hot sparge water (around 78°C) will help in reducing wort viscosity, but the temperature should not be too high otherwise extraction of tannins from grain residues may occur. In practice, to achieve an acceptable run-off rate, the combination of relatively high viscosity and low bed permeability limits the depth of the grain bed to between 30 and 50 cm.

2.3 Boiling

After lautering, the sweet wort is transferred to the kettle or "copper," in reference to the traditional material used in fabricating the kettle. Modern kettles are almost universally made from stainless steel and include the ability to whirlpool the wort at the end of the boil to separate and compact the *trub*. The boiling step serves multiple purposes:

1. To sterilize the wort prior to fermentation
2. To remove unwanted volatile components
3. To extract flavor and aroma compounds from hops
4. To isomerize hop acids
5. To increase color through Maillard reactions

Figure 2.6. A traditional brewhouse with multistory vessels made from copper. The shape of the domed top affects the retention of volatiles during heating.

6. To denature and precipitate proteins, tannins and minerals to form the trub (referred to as the "hot break")
7. To separate the trub by whirlpooling (in some designs)

The kettle can be heated in different ways: with the use of steam jackets, direct firing by flame or by recirculation of the wort through an external heat exchanger. When the wort is heated to boiling temperature, the hot surface of the vessel or heat exchanger can eventually become fouled by the production of a scale referred to as "beer stone," a precipitate composed of calcium oxalate. The design of the kettle differs, often from brewery to brewery, and can affect the characteristics of the beer. The domed shape of the top and size of the exhaust stack can both affect the retention of volatile components in the wort while the boil proceeds.

Because boiling serves multiple purposes, the time period used for the boil may vary depending on the situation, as determined by the composition of the wort and the type and quantity of hops to be used. Boil times can be between 60 minutes to several hours and could include a rest period at the end prior to whirlpooling and cooling. Typically, the bittering hop varieties with high alpha-acid content are added early in the boil to provide sufficient time for isomerization to occur. The aromatic varieties, such as Cascade or Saaz, are added near the end to allow enough time for extraction but not so much time as to distill off the aroma. More information on different hop varieties and the chemical changes that occur during boiling can be found in Chapter 6.

In addition to the loss of aromatics, boiling also results in a reduction in the wort volume due to loss of water vapor out the vent stack. This loss will cause an increase in the sugar concentration, as measured by specific gravity (SG) or degrees Plato (°P), and must be accounted for when targeting a desired OG prior to fermentation. As an estimate, losses are usually between 2 and 5% depending on the design of the kettle and the boil time. Thus, with a target OG of 1.048 or 12°P, the pre-boil SG should be between 1.046 and 1.047.

At the conclusion of the boil, the hot wort contains significant amounts of suspended solids, consisting of hop and grain residues, precipitated proteins and minerals. These solids are collectively referred to as *trub* and should be removed as much as possible from the wort before being transferred to the fermenter. The most common method utilizes a whirlpool effect, a concept first used by Ranulph Hudston while working at the Molson Brewery in Montreal in 1960. He had modified the traditional hot wort receiver, which separated the trub using simple sedimentation, by changing the wort entry point to a tangential connection. By pumping the wort out of the bottom of the kettle, offset from the center, and into the tangential entry, the whirlpool motion was induced, resulting in the migration of the trub to the center of the vortex. After whirlpooling for 20–30 minutes, the clarified wort could be transferred for cooling and aeration, leaving the trub behind. Separate vessels for whirlpooling are also sometimes used today, especially when the kettles are very large.

2.4 Cooling and Aeration

After the boil has been completed and the trub separated, the hot wort must be cooled and aerated prior to entering the fermenter. This

Figure 2.7.   A plate heat exchanger (left) and 0.45 μm air filter and pressure regulator (right) for cooling and then aerating the wort during transfer from the kettle to the fermenter.

is usually accomplished by pumping the wort through a plate-type heat exchanger, cooled by chilled water or glycol solution, followed by in-line aeration by filtered air through a sintered aeration device.

The cooling capacity of a specific heat exchanger is based on the total surface area of its plates and the inlet temperature and flow rate of the cooling medium. The cooling capacity must be matched to the flow rate of the wort during transfer and the fermentation temperature requirements. The use of a glycol solution for a cooling medium can reduce the inlet temperature below the freezing point of water and therefore reduces the size of the heat exchanger needed.

The extent of wort aeration to be used is dependent on the requirements of the yeast culture that will be added to the fermenter subsequent to the cooling and transfer. Generally, it is a good idea not to over-aerate (as can happen when bottled oxygen is used) as this can cause an oxidative-stress reaction in the yeast and also start staling reactions in the wort that will ultimately reduce the shelf-life of the beer. If non-enriched air is used then saturation of the wort to an oxygen concentration between 8 and –10 mg/L (oxygen solubility is inversely correlated with temperature) is normally sufficient to ensure a rapid start to the fermentation. Also, if the yeast has been freshly propagated under aerobic conditions or has been re-activated aerobically after cropping from a previous batch, aeration may not be needed at all.

2.5 Topics for Discussion

1. What are the advantages and disadvantages of a 3-vessel brew-house in comparison to a 2-vessel system?
2. What issues need to be considered when deciding on the use and selection of adjuncts?
3. Other than the traditional method of adding hops to the kettle, what other strategies could be employed to flavor the beer?
4. If the run-off rate from the lauter tun is unacceptable, what strategies could be used to reduce the time required?
5. If enzymes are added to an all-grain mash to improve starch or protein degradation, what impact could this have on the subsequent fermentation and characteristics of the final beer?

2.6 Further Reading

1. Boulton C. *Encyclopaedia of Brewing*. Chichester, West Sussex: John Wiley & Sons Ltd.; 2013.
2. Handbook of Brewing. In: Hardwick W, ed. New York, NY: Marcel Dekker Inc.; 1995.
3. Lewis M, Young T. *Brewing*. London, UK: Chapman & Hall Ltd.; 1995.
4. *The Practical Brewer*. 3rd ed. Wauwatosa, WI: Master Brewers Association of the Americas; 1999.

Chapter 3

Yeast and Fermentation

3.1 Origins of Brewing Yeast

In many ways the current strains of brewing yeast are ideally suited for their role in the brewing process, especially for mass-marketed beer styles such as light lagers that appeal to a broad cross section of consumers. These desirable characteristics include tolerance to elevated levels of both ethanol and sugar, high rate of fermentation and yield of ethanol from different sugars under a variety of environmental conditions, while still retaining the ability to maintain sufficient viability and vitality to be reused for multiple generations. But how did it happen that the characteristics that the brewer so desires are manifest by species of *Saccharomyces*? To begin to answer this question, scientists believe that several distinct genetic events occurred between 80 and 100 million years ago that provided *Saccharomyces cerevisiae* with a very significant competitive advantage. These events included duplication of their entire genome, which resulted in the formation of several new gene pairs involved in ethanol metabolism. These genes gave *S. cerevisiae* the ability to produce ethanol even in the presence of oxygen because of at least partial repression of respiration by the presence of sugars such as glucose. This is referred to as the Crabtree-positive effect. The accumulation of ethanol in the environment of the yeast proved to be toxic to many other microbes that competed for nutrients. Furthermore, additional genes evolved, coding for novel forms of alcohol dehydrogenase enzymes, allowing *S. cerevisiae*

33

to utilize the accumulated ethanol as a carbon source for its own growth. The combination of these genetic traits gave *S. cerevisiae* an ability to alter its environment in a manner that provided a toxic situation for its competitors without having to pay a high metabolic price.

The ubiquitous presence of *Saccharomyces sp.* in the environment ensured that the fermentation process would eventually be discovered by early civilizations. All that was needed was some type of vessel, such as a clay pot, containing a source of fermentable sugar with an adequate level of moisture to support yeast metabolism. Early fermentation would have been based on multiple species of microbes, each contributing something to the success of the brew. However, a successful result was certainly not guaranteed as not all microbes were desirable, many contributing various acids, unpleasant flavors and odors. Also, the level of ethanol produced by the yeast was likely to have been quite low based on today's standards. So, over many centuries, the process was refined in order to increase the quality and consistency of the desired end product. Unbeknownst to the practitioners of the time, as the brewing process was refined, so were the various populations of yeast that were inherent to the success of their efforts. This gradual improvement in the characteristics of brewing yeast can be referred to as a process of yeast domestication.

The concept of domesticating wild animals to better serve the purposes of society is well known, but historically has referred to a conscious effort with animals that are easily observable. Over many generations, animals are selected and bred to encourage the development of specific, more desirable characteristics. While some characteristics become improved upon, other characteristics are lost and generally the animal is no longer equipped to compete and survive in the wild. The genetic effects of domestication may include genome decay, polyploidy/aneuploidy, chromosomal rearrangements, gene duplications and new phenotypic behaviors.

For the most part, the domestication of yeast occurred without the knowledge of the brewer and was, therefore, an unconscious process. Its success was based on how the brewing process evolved empirically in concert with the metabolic behavior of the yeast. Even though the role of yeast was not confirmed scientifically until the

19th century, the presence of yeast had been indirectly inferred based on how the brewing process had evolved to provide more rigorous control over fermentation conditions, including control of the "backslop" or reuse of solids from previous batches.

Milestones in the genetic changes that accompanied the domestication of industrial ale yeasts have been elucidated by a group in Belgium that sequenced and phenotyped 157 strains of *S. cerevisiae*, obtained from various industrial sources, but primarily brewing environments. From their analyses, they divided current industrial strains into five categories or sublineages that showed evidence of domestication based on significant differences from wild strains both genetically and phenotypically. The metabolic differences that were considered included: flavor production, flocculation, ethanol production, environmental and nutrient stress tolerance, maltose and maltotriose fermentation capacity and 4-vinyl guaiacol (4-VG) production.

The five categories or clades of yeast were referred to as: Wine, Beer1, Beer2, Asia (sake) and Mixed (including bread). The evolutionary divergence of the yeast strains in these five categories was correlated with both the industry within which they were being used and also their geographical origin. Both Beer1 and Beer2 yeasts showed strong evidence of domestication based on several genetic differences compared to wild types of *S. cerevisiae*, including:

- Mutation and duplication of the *MAL* genes that affect maltose utilization
- Mutations in *PAD1* and *FDC1* that affect the production of the off-flavor 4-VG
- Aneuploidy/polyploidy and loss of sexual function

Although not seen in the Beer2 category, within the Beer1 category there were three distinct subcategories (or subclades) based on geography: Britain, the United States and Belgium/Germany. The approximate time when the genetic divergence of these various strains occurred, required calculations based on a number of assumptions. The United States subclade from the Beer1 strains, was most closely related to the subclade from Britain and since it is known that America was colonized by the British in the early 17th century, this gave an estimate of the time required for this divergence to occur. Furthermore, assuming that a population of yeast in a brewing environment undergoes approximately 150 asexual doublings per year, an average mutation rate of $1.61–1.73 \times 10^{-08}$ base pairs per generation was determined. This number, along with the magnitude of genetic diversity between categories, was then used to estimate the dates at which various divergent events occurred.

The last common ancestor for Beer1 yeasts is estimated to have disappeared between 1573 and 1604, while Beer2 yeasts probably diverged from wine yeast between 1645 and 1671. However, prior to these major divergent events there is evidence for genetic decay, loss of sexual reproduction and reduced tolerance to stress, as well as the development of positive traits such as increased ethanol tolerance, maltotriose utilization and reduction in off-flavors. So, it is likely that domestication of ale yeast from wild types had already been in progress for some time.

Compared to the history of *S. cerevisiae*, lager yeasts are relative newcomers and have only been on the brewing scene since sometime in the 15th century, although now, over 90% of the beer produced worldwide is considered to be lager. Unlike *S. cerevisiae*, the lager yeasts *S. pastorianus* (or *S. carlsbergensis*) are not naturally occurring but are hybrids, likely bred in the brewing environment. Their precise origin is not known but because of their tolerance to cold, it was thought that they were related to the species *S. uvarum* or *S. bayanus*. Recently, genetic typing has revealed that lager yeast is an interspecies hybrid of *S. cerevisiae* and *S. eubayanus*. The latter species was originally isolated from a tree in Patagonia and more recently has also been isolated in parts of the United States, but not yet in Europe, so, it is unclear where the mating(s) may have occurred. Lager yeasts are of two types, Frohberg and Saaz, classified based on the relative proportions of genetic material obtained from each parent. Saaz strains are closer to their *S. eubayanus* parent with two compliments of the *eubayanus* genes and only one copy of the *cerevisiae* genes. This is also reflected in their metabolism as they cannot ferment maltose and maltotriose as well as *cerevisiae* strains but are more cold tolerant.

In contrast, the Frohberg strains have two copies of genes from each parent but resemble *S. cerevisiae* more closely in fermentation performance and are more commonly used as industrial strains than Saaz. There is not yet general agreement as to whether the hybridization event was a single occurrence and the two types diverged from there, or whether there were two separate events and Frohberg and Saaz have somewhat different parentage. It is generally considered that domestication of lager yeast began after the hybridization occurred, as *S. eubayanus* was not widely used as a brewing yeast prior to its mating with *S. cerevisiae*. However, if there were two mating events, it is likely that both of the *S. cerevisiae* parents had already began to diverge from each other, based on previous domestication in different brewing environments.

3.2 Yeast Reproductive Cell Cycle

Yeast are unicellular eukaryotic microorganisms with significantly greater structurally and metabolic complexity than found in prokaryotic microbes such as bacteria. Also, an important distinguishing characteristic of brewing yeast, is that although oxygen has a dramatic impact on their metabolism, they are able to grow and produce ethanol, under either aerobic or anaerobic conditions and are, therefore, referred to as *facultative* microbes. The distinguishing features of eukaryotic cells relate to the various specialized organelles within the cellular matrix:

- Nucleus (enclosing the DNA)
- Mitochondria (generating ATP from respiration)
- Vacuoles (regulation of pH and ion flow)
- Endoplasmic reticulum (protein synthesis and lipid metabolism)
- Spindle pole body (cell cycle regulation and mitosis)
- Plasma membrane (regulates gradients and biosynthesis)
- Cell wall (provides structural integrity and protection)

Like all eukaryotic cells, *Saccharomyces sp.* reproduce by completing a programmed sequence of metabolic events that define what is

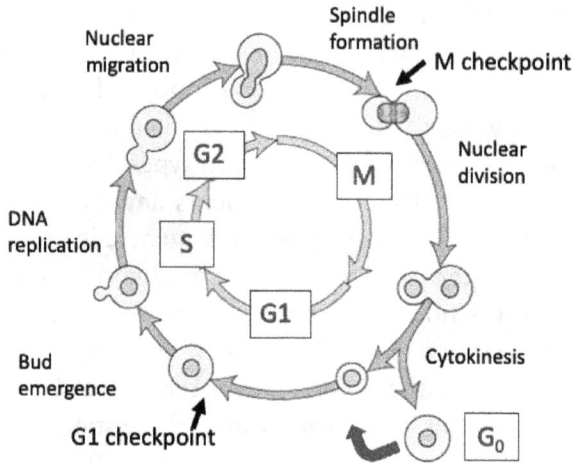

Figure 3.1.   Phases of the budding yeast cell cycle with progression from G1, through S, G2 and M. G_O phase refers to a resting or dormant cell.

called the *cell cycle*. However, unlike most eukaryotes, brewing yeast reproduce by budding rather than fission, resulting in an asymmetry between the mother and daughter cells due to unequal distribution of cellular material.

As illustrated in Figure 3.1, the yeast cell cycle can be considered as having a start point at the time of bud separation or birth of the daughter cell. The first phase is referred to as Gap1 (G1) and represents the time required for the cell to grow and reach a critical size before it can progress past the G1 checkpoint. The time for this to occur will be based on the yeast strain, its growth environment and its previous history. At the time of bud separation, daughter cells are typically smaller than the mother cell and require additional time to reach the critical size. Mother cells may have been exposed to the specific growth environment for several previous generations or passages through the cell cycle and have adjusted or adapted their metabolism to make optimal use of the available nutrients. At the time of bud separation, if the environment is unable to support growth, then both daughter and mother cells may enter a resting state, referred to as the G_O phase. No growth occurs during G_O but the cell remains viable and the G1 phase will resume if and when the appropriate environmental conditions are reestablished. If the critical size has been attained and no DNA damage is detected, the cell will pass the G1 checkpoint and begin formation of a new bud followed by DNA synthesis and replication in the S phase. By the time this has occurred and the cell has entered Gap2 (G2), the bud will have increased in size and the nucleus will have migrated to the location of bud attachment. As the cell enters the final stage, metaphase or M phase, it must pass a second checkpoint before making a final commitment to replication. If DNA has not been duplicated correctly, damage is detected or if the chromosomes are misaligned on the spindle assembly, the cell cycle will be arrested. If all is well, the chromosomes will be segregated between mother and daughter, forming two separate nuclei, followed by cytokinesis and the birth of the daughter cell. As the bud separates, a permanent scar composed of chitin is left on the surface of the mother cell. The number of scars present on a yeast cell is therefore an indication of its replicative age. As the yeast cell ages, certain metabolic changes occur, eventually leading to an inability to complete additional cycles, referred to as senescence. The number of cycles that a eukaryotic cell can complete during its replicative life span is referred to as its *Hayflick limit*. For *S. cerevisiae* this limit may vary greatly, ranging between 8 and 25 depending on both environmental and genetic factors.

Internal control of the yeast cell cycle and passage through the two checkpoints is accomplished using a class of proteins called *cyclins*. As the cycle progresses, nine different cyclins combine with a cyclin-dependent kinase (Cdk) to form dimers that are responsible for various cellular functions. Regulation of these functions is accomplished by activation and inactivation of these Cdk/cyclin dimers using several different mechanisms to achieve both cyclin synthesis and degradation.

3.3 Propagation

Cells require energy to grow and reproduce and like all eukaryotes, yeast rely on glycolysis as a mechanism for obtaining energy from sugars. Figure 3.2 shows how the end product of glycolysis, pyruvate, is either fermented or converted to Acetyl-CoA during respiration.

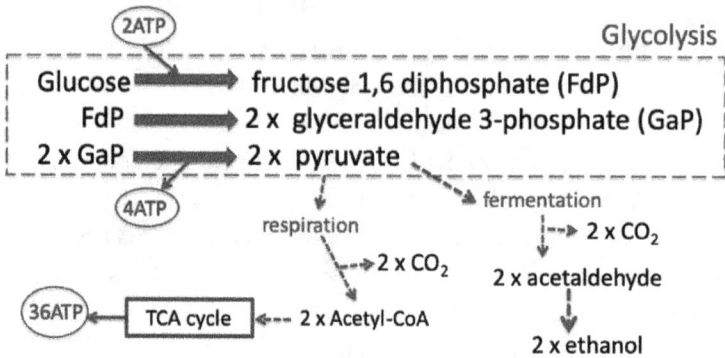

Figure 3.2. The three main stages of glycolysis consist of 10 steps, generating a net of 2 ATP molecules per molecule of glucose. The pyruvate is either fermented to ethanol and CO_2 or is converted to Acetyl-CoA and CO_2 before entering the TCA cycle under aerobic conditions to generate up to an additional 36 molecules of ATP.

Since brewing yeast are crabtree-positive, even under fully aerobic conditions, some pyruvate will be fermented to ethanol, but this can be reassimilated once the glucose from the environment has been exhausted. Although the cells prefer glucose as the primary source of energy, unless adjuncts have been used in the mash, brewer's wort typically contains only small quantities of the monosaccharides glucose and fructose. The primary sugar is the disaccharide maltose, a product of the activity of amylase enzymes originating in the malted barley. Both maltose and the trisaccharide maltotriose enter cells by an active transport mechanism using permeases, requiring the cells to expend energy, unlike glucose and fructose which enter the cell by passive diffusion. Once inside the cell, both maltose and maltotriose are acted on by the α-glucosidase enzymes, which cleave the α-1,4 bonds to produce glucose monomers. The presence of glucose (and also sucrose when sugar adjuncts are used) can result in a reduced rate of transport and utilization of both maltose and maltotriose, a phenomenon more pronounced in strains of ale yeast than lager.

Growth of yeast for the purpose of generating new cells, for example prior to starting a new beer fermentation, is best accomplished

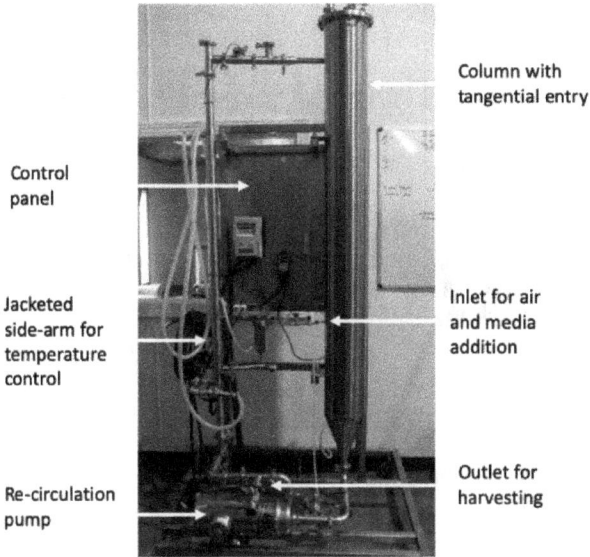

Figure 3.3.   A cyclone column bioreactor for use as a yeast propagator. Oxygen transfer is accomplished across the swirling liquid film inside the column and from entrained bubbles in the recirculation loop.

using a specially designed propagator or *bioreactor.* Unlike a fermenter, a propagator supports rapid growth of the yeast under well-mixed aerobic conditions. Mixing is accomplished using either a mechanical agitator or pumped recirculation as used in the cyclone column design pictured in Figure 3.3.

Air is introduced into the turbulent liquid medium in sufficient quantities to provide an adequate level of dissolved oxygen to support aerobic growth of the desired concentration of cells. Both the aeration and agitation rates will have large effects on the rate of oxygen transfer into the liquid medium and so, the oxygen concentration should be monitored and controlled based on the use of an on-line dissolved oxygen probe.

The growth kinetics of a cell population operated as a batch system, in which all nutrients are provided at time zero, is illustrated in Figure 3.4. The growth can be described by four phases: lag,

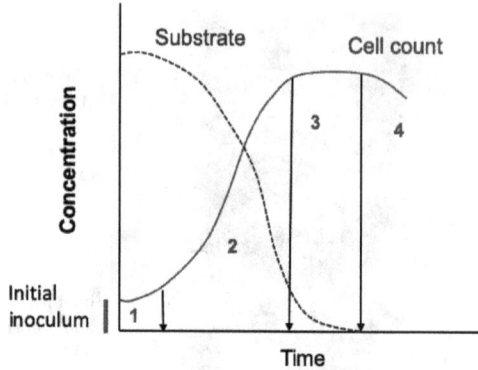

Figure 3.4.   The four stages of the batch growth cycle: 1-lag, 2-exponential, 3-stationary and 4-death. The cell population enters stationary phase as the growth substrate is depleted.

exponential, stationary and death. These phases are not to be confused with the underlying cell cycle, and are purely a result of the manner in which nutrients are supplied to the cells. The lag phase is characterized by very slow growth as the cells adapt to the new environment. It can be of variable duration based on the metabolic state of the inoculum, quantity of the inoculum in comparison to the volume of the propagator and characteristics of the growth environment such as nutrient levels, pH and temperature. The second phase is when the cells are growing at their maximum rate and cell number increases exponentially. For a specific strain of yeast growing in a controlled environment, the growth rate should be constant during this phase and can be represented by the *specific growth rate*, μ. The value for μ can be calculated if the cell concentration is obtained from at least two time points during the exponential phase:

$$\mu = \ln(N_2/N_1)/(t_2 - t_1)$$

where N_x represents cell concentration at time point t_x. The units for specific growth rate are inverse time, typically per hour, and is related to the cell doubling time by the following equation:

$$t_d = 0.693/\mu$$

The cell doubling time should be consistent for each batch unless the strain of yeast has been changed or the environment altered. Under aerobic conditions brewing yeast typically have doubling times ranging between 3 and 6 hours. As the cell population increases, nutrients are consumed from the growth medium and eventually the growth slows and stops as an essential nutrient is no longer available and becomes the limiting factor. Availability of sugar is often the growth-limiting factor, so, adjusting the initial sugar concentration in the growth medium can be used to control the final cell count. There are, however, limits as to how much sugar can be used, as excessive concentrations can result in inhibition of growth. The stationary phase is characterized by a constant cell count as there is no appreciable growth due to lack of at least one essential nutrient. The cells will retain viability, living off of stored reserves such as glycogen. However, eventually viability will be lost and cell count will decrease as the cells enter the death phase.

The ideal time to harvest cells for use in a beer fermentation is late in the exponential phase while cells are still actively growing. This will help ensure that the subsequent fermentation will start quickly with a minimum lag time.

Batch culture, although simple to operate, is not particularly effective for generating high concentrations of cells due to the limitations on starting nutrient concentrations. An alternative approach is to feed the cells with additional nutrients before they are fully depleted, a strategy referred to as *fed-batch*. This effectively extends the exponential growth phase and results in an increase in the final cell count as illustrated in Figure 3.5. The limitations of the fed-batch strategy are that at high cell concentrations other issues such as mixing intensity, oxygen transfer rate or cooling capacity of the propagator may not be sufficient to maintain optimal performance. Also, there will be physical limitations as to the extra working volume that is needed for addition of the supplemental nutrients.

Even by implementing a supplemental feeding strategy, the quantity of cells that can be grown in a single batch is limited by the volume of the propagator. If additional cells are needed in order to provide sufficient starting cell count in the beer fermentation, then

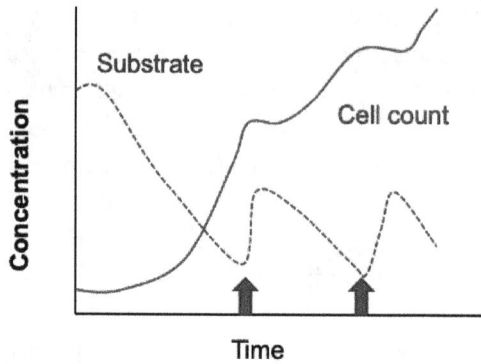

Figure 3.5. Fed-batch feeding strategy with the arrows representing the time for addition of supplemental substrate.

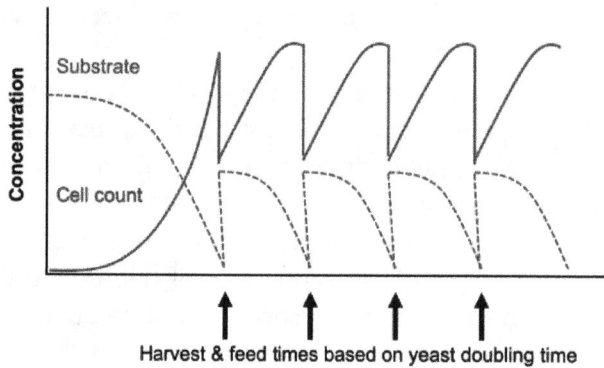

Figure 3.6. Semi-continuous method for propagating additional quantities of yeast for use in large fermentations. An automated system based on dissolved oxygen can be used for determining precise times for initiating a harvest/feed cycle.

a semicontinuous strategy can be employed as illustrated in Figure 3.6. This strategy involves periodic harvesting and feeding of the cells at equal volume, ideally one half of the total working volume. If the timing of each harvest/feed cycle is based on the doubling time of the cells, then the method is referred to as *continuous phasing*.

Furthermore, if an automated control system is used to determine when the cell doubling has occurred and the precise time to initiate the harvest/feed cycle, the process is referred to as a *self-cycling fermentation*, because the metabolism of the cells trigger the harvest/feed cycle for each succeeding generation. If employed for at least 3 cycles, both continuous phasing and self-cycling fermentation can result in a significant degree of synchrony with respect to cell division, leading to greater metabolic uniformity in the cell population.

Various media formulations can be used for yeast propagation, ranging from a fully defined recipe based on purified ingredients to brewer's wort. In any case, the concentrations of essential nutrients will determine the final cell count that can be attained. For example, brewer's wort at 12° Plato or dried malt extract (DME) at approximately 150 gL^{-1}, should provide sufficient nutrients to obtain between 5×10^8 to 1×10^9 cells per mL. These complex media formulations are preferable over defined ingredients because they most closely resemble the nutrient environment the cells will encounter in the beer fermentation.

3.4 Beer Fermentation

Once the sweet wort has been cooled, aerated and transferred from the brewhouse into the fermentation suite, the heart of the brewing process can begin. It is a truism to say that "Brewers make wort but yeast make beer." Figure 3.7 illustrates the progress of a typical lager fermentation of a 12° Plato wort. Note that most of the cell growth occurs during the first 72 hours while ethanol production continues until reaching the final gravity. As the availability of sugars decreases in the last stage, cell concentration drops due to an increase in flocculation and settling of the suspended yeast cells.

The most important aspect to beer fermentation is the ability of the brewer to control the consistency of the process and associated quality of the end product. To achieve an acceptable level of process

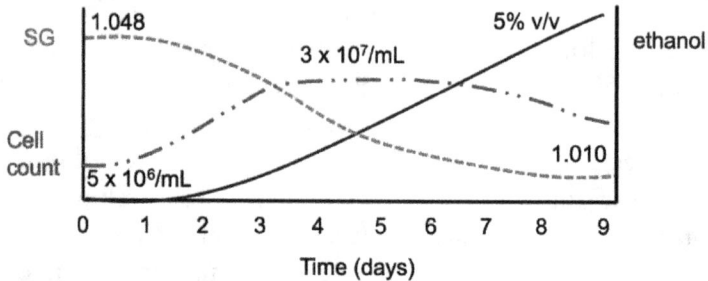

Figure 3.7.   Progress of a typical lager fermentation showing changes in the concentration of yeast, sugars (by specific gravity) and ethanol.

control, batch after batch, requires careful attention to the following parameters:

1. Wort composition
2. Yeast quantity (pitching rate)
3. Yeast quality (purity and brewing fitness)
4. Fermentation temperature and time

Wort Composition

The composition of the wort is quite complex, containing many different compounds that can influence yeast metabolism during fermentation, including various sugars, sources of nitrogen, lipids, polyphenols, esters, free fatty acids, aliphatic acids and trace elements such as vitamins, coenzymes and minerals.

Not all of the wort components contribute to the fermentation but, fortunately, brewing yeast strains are generally well adapted to the brewing environment and an all-grain derived wort is likely to contain all that is necessary for the yeast to complete a successful fermentation. However, there can be major differences between wort, depending on the recipe and mashing protocol used.

In addition to the type and quantity of sugars, a critical factor affecting yeast fermentation performance is the availability of nitrogen, mostly in the form of amino acids. Although yeast are able to

Table 3.1. The approximate fractions of the major components of a typical all-grain wort. Use of adjuncts to replace or supplement barley malt, can affect these percentages.

Component	Fraction (% of dry weight)
Total Carbohydrates	90–92
Monosaccharides (glucose, fructose, sucrose)	15
Disaccharides (maltose)	40
Trisaccharides (maltotriose)	20
Dextrins	15
Total Nitrogen	3.5–6
Amino acids (FAN)	1.6
Peptides	1.6
Protein	1.4
Other	0.7
Total Minerals	1.5–2
Vitamins, Lipids and Phenols	0.15–0.25

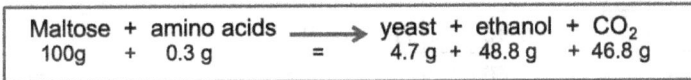

Maltose + amino acids $\longrightarrow$ yeast + ethanol + CO_2
100g + 0.3 g = 4.7 g + 48.8 g + 46.8 g

Figure 3.8. This simplified materials balance for a beer fermentation does not include other sugars such as glucose, fructose and maltotriose and products of yeast metabolism such as glycerol, flavor and aroma compounds.

synthesize many of the amino acids necessary for growth, depending on the sugar concentration, a successful fermentation requires a certain minimum level of amino acids to be present in the wort. A simplified materials balance is presented in Figure 3.8.

Although the quantity of nitrogen in the wort is often represented by measurement of the *free amino nitrogen* (FAN), this oversimplifies the impact of different amino acids, especially on flavor development. Amino acids have been grouped into four categories (see Table 3.2) reflecting how quickly they are utilized by brewing yeast. Group B are

Table 3.2. The four groupings of amino acids as related to how quickly they are absorbed from the wort.

Group A fast uptake	Group B intermediate	Group C slow	Group D minimal
Arginine Asparagine Aspartic acid Glutamine Glutamic acid Lysine Serine Threonine	Histidine Isoleucine Leucine Methionine Valine	Alanine Ammonia Glycine Phenylalanine Tryptophan Tyrosine	Proline

Table 3.3. The major inorganic cations found in brewer's wort and the potential effects on yeast and beer quality.

Inorganic ion	Concentration range (mg/L)	Potential effects on fermentation and beer quality
Calcium	40–100	Enhances enzyme activity, reduces haze, improves yeast growth and flocculation
Magnesium	10–15	High levels can adversely affect taste
Potassium	300–700	Important in determining alkalinity, aids in phosphate utilization
Sodium	10–150	Improved mouthfeel
Manganese	0.1–0.2	Required for yeast growth, improves protein solubility, may affect color
Iron	0.08–0.4	Reduces some enzyme activity, required for yeast growth, catalyzes oxidation
Copper	0.1–0.5	Can reduce haze, catalyzes oxidation
Zinc	0.07–0.12	Low levels enhance yeast growth, can cause haze and beer oxidation

especially important in flavor production as described in detail in Section 3.5. Other important characteristics of the wort include the pH and inorganic ion composition. Both of these parameters can affect yeast metabolism and the final quality of the beer.

Inorganic ions such as zinc and calcium need to be present in the wort in sufficient quantity to promote a healthy yeast population, but not in excessive amounts that may adversely affect beer quality.

Pitching Rate

The pitching rate defines the initial concentration of yeast cells in the fermenter and has important consequences for the progress of the fermentation and the quality of the beer produced. The ideal pitching rate for any given fermentation is influenced by several factors and needs to be arrived at by the brewer based on their experience with a specific recipe, yeast strain and the yeast management strategy employed. However, to give a good starting point there are general guidelines that can be followed. The main factors that need to be considered are the concentration of sugars in the wort (as reflected by the original gravity), whether it is a lager or ale yeast and the current state of "brewing fitness" in the yeast population. In general, for ale yeast it is recommended to use between 0.4 to 1.0×10^6 cells per mL per °Plato of the wort. Lager yeast are usually pitched at a higher rate, between 1.0 to 1.65×10^6 cells per mL per °Plato, due to the slower metabolism of *S. pastorianus* fermenting at a lower temperature than *S. cerevisiae*. So, with a typical 12 °Plato wort an ale pitch rate would be between 4.8 and 12 million cells per mL and a lager between 12 and 20 million cells per mL. To achieve these pitching rates requires knowledge of the cell concentration in whatever yeast is to be used, as the concentration can vary widely depending on the source.

Table 3.4. The calculated pitch quantities are for an ale yeast based on a 12 °Plato wort and a pitching rate of 18×10^6 cells per mL. Higher gravity wort and use of lager yeast would require an additional quantity.

Source of yeast	Average cell count	Approx. pitch quantity (per bbl of fermentation)
Active dry	2×10^{10} per gram	0.1 kg
Harvested slurry	3×10^9 per mL	0.7 liters
Propagated slurry	1×10^9 per mL	2.1 liters

Yeast Quality

A population of yeast contains billions of single cell organisms, some in good health (referred to as being *vital*), others in a damaged state and others that are dead. The relative distribution of these subpopulations defines what is called the *brewing fitness* of the culture. Many factors can affect the brewing fitness and are a product of the yeast management strategy employed in the brewery.

A well designed and executed yeast management strategy will maintain the purity of the culture, a high percentage of viable cells and a metabolic condition that is well-suited to beer fermentation. Yeast management involves a series of procedures that may include propagation, pitching, harvesting, washing, storage and re-pitching. How these procedures are carried out, combined with the characteristics of the specific strain, will determine the overall brewing fitness of the yeast population, as defined in Figure 3.9. Propagation and pitching procedures were described in previous sections of this chapter. Harvesting or *cropping* of the yeast in a concentrated slurry is performed at the end of primary fermentation. Lager yeast is considered as "bottom-fermenting" yeast and is readily harvested from the bottom outlet of the primary fermenter. The settling of yeast still in suspension can be enhanced by implementing a *cold-crash* at 4°C for 24 hours prior to harvesting. Ale yeast is considered to be "top-fermenting" and tends to form a yeast blanket at the top of the fermenter, although some fraction of the population also settles to the bottom. This fractionation of the population may be related to the

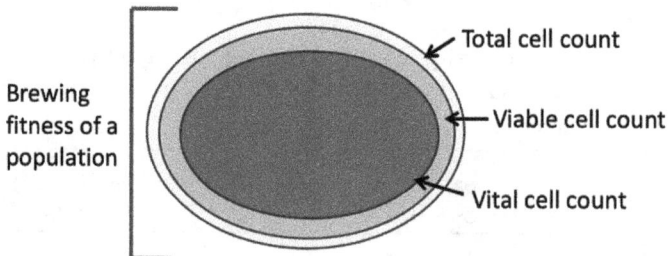

Figure 3.9.   The brewing fitness of the yeast population is based on the proportion of viable and vital cells compared to the total cell count.

size, age and flocculation differences between the individual cells. If the harvested yeast is to be re-pitched, it is advisable to harvest both fractions in order to maintain a balanced population. The yeast blanket can be removed from the bottom outlet once the ale has been transferred from the primary fermenter to the maturation vessel.

Prior to re-pitching the concentrated slurry of harvested yeast should be evaluated with regards to its purity (freedom from bacterial contamination), plus total and viable cell counts. Even if there is no evidence of bacterial contamination some brewers perform a wash step, which involves subjecting the yeast to a phosphoric acid solution at a pH between 2 and 2.2 at a temperature less than 5°C while gently agitating for a maximum of 2 hours. This acidic treatment should kill any contaminating bacteria but may also reduce the yeast viability and vitality, so, should be carried out with caution. Alternatively, if bacterial contamination occurs the yeast can be discarded and a fresh culture can be purchased or propagated and the source of contamination identified and rectified. Following good sanitation practice should prevent most contamination issues from occurring.

Yeast cell counts can be performed using a variety of methods, which vary with respect to both convenience and accuracy. Some methods will provide only a count of total cells, while others provide both total and viable and, in some cases, damaged or non-vital cells are not counted as being viable. A summary of the most common methods is presented in Table 3.5. Of the manual methods, agar plate

Table 3.5. The most common methods for determining yeast cell counts. Depending on the level of cell vitality, each method may provide differing results for the viable cell count.

Method	Equipment used	Type	Data obtained
Agar plating	Hood, incubator	Manual	Viable count
Capacitance	Probe	Automatic	Viable-vital count
Light absorbance	Spectrophotometer	Manual	Total count
IR absorbance	Probe	Automatic	Total count
Staining + visual exam	Microscope	Manual	Total + viable count
Staining + image analysis	Cell counter	Automatic	Total + viable count

counting and staining with methylene blue followed by microscopic examination are the most frequently employed.

Agar Plating

Petri plates or dishes are prepared with a sterile formulation of various nutrients and semi-solid agar to support the growth of viable cells. In the spread-plate method, a sample of the yeast culture is diluted appropriately in either sterile water or buffer solution to provide between 30 and 300 viable cells, or *colony-forming units* (CFUs) per mL. One mL of the diluted sample is then spread evenly over the surface of the agar, incubated for 24–48 hours at controlled temperature and then counted for the number of separate colonies that have formed. If the initial concentration of viable cells is completely unknown, then a series of different dilutions must be prepared to ensure that the number of colonies formed on the plate is in the countable range. The cell count present in the original sample is then calculated based on the dilution factor used. To ensure a measure of statistical validity, usually plate counts are performed in triplicate and the results averaged. An alternative procedure, which uses fewer plates and smaller sample volumes, is the "tracking" method. In this procedure, square agar plates with six separate lanes or tracks are used (see Figure 3.10). Twenty

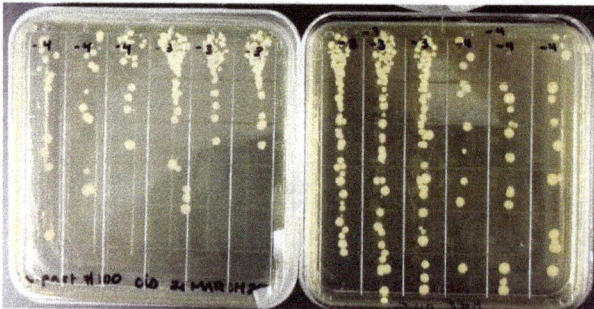

Figure 3.10.   Two agar plates on which samples of yeast at various dilutions have been placed on separate tracks for determining viable cell counts.

microliters of sample are placed at the start of each track and then the plates are turned vertically, so that the samples run down each track, leaving individual cells deposited and separated on the agar surface. After incubation, individual colonies can be counted as with the traditional method, however, with six separate tracks, multiple dilutions can be tested on a single plate. Although the agar plating method is overall the best indicator of viable cell count, it does have some drawbacks. It requires considerable manual work for preparation of plates, great care must be taken in preparing accurate dilutions and the test results are not available for at least 24 hours. Furthermore, non-vital cells may or may not be capable of forming colonies, depending on the nature of damage they have sustained and their ability to recover.

Staining and Microscopy

An alternative method for determining both total and viable cell counts is based on staining a sample of yeast cells in suspension with a dye (typically methylene blue or methylene violet) and then a known volume of the stained cells is examined microscopically to determine cell number. Healthy, viable cells will reject the dye from the cell membrane and remain colorless, while non-viable cells or cells with low vitality will appear blue. Thus, both total and viable counts can be obtained from a single sample. This method can be employed using a manual procedure that is based on the use of a glass slide with an etched grid (hemocytometer) and a microscope, or an automated instrument (such as the Nexcelom Vision Cellometer) that incorporates both the optics and image analysis software to count large numbers of both stained and non-stained cells much more quickly than can be accomplished manually. As with the agar plating methods, it is necessary to test samples with an appropriate dilution factor to ensure that the density of cells is not too high or too low to prevent accurate enumeration. Also, the accuracy of the viability counts is dependent on the sensitivity of color recognition, whether manual or automatic, and is generally not recommended for yeast populations that possess low levels of viability.

In summary, yeast quality as reflected by its brewing fitness is a complex concept that cannot be measured by a single test. It is most important that the brewer becomes familiar with the characteristics of the specific yeast strain being employed and to adjust the yeast management strategy to account for the metabolic differences between strains. Although most brewing strains are quite robust, to obtain the most consistent fermentation performance and the highest quality end product, the brewer must pay constant attention to the well-being of the yeast culture.

3.5 Yeast and Flavor

The complex flavor and aroma of beer is a result of the intermingling of dozens, if not hundreds, of different chemical compounds, the majority of which are produced by the yeast during fermentation as metabolic byproducts. Some of these compounds are desirable and within certain ranges help to define the style of beer, while others are considered off-flavors and should be minimized. The key compounds are classified as higher alcohols (fusel alcohols), esters, vicinal diketones (VDKs) and aldehydes, which are primarily associated with beer staling. Table 3.6 lists some common flavor and aroma compounds

Table 3.6. Some of the important flavor and aroma compounds produced by yeast during fermentation. The taste threshold varies based on the individual and the specific beer matrix.

Component	Type	Effects on beer	Taste threshold (mg/L)
Isoamyl alcohol	Fusel alcohol	Alcohol, banana	50–65
Phenylethyl alcohol	Fusel alcohol	Alcohol, flower, honey	40
Phenylethyl acetate	Ester	Rose, honey apple	0.2–3.8
Ethyl acetate	Ester	Solvent, pear, sweet	25–30
Isoamyl acetate	Ester	Fruity, banana	1.2–2.0
Ethyl hexanoate	Ester	Apple, anise	0.2–0.23
Butanedione (diacetyl)	VDK	Butter	0.05–0.15

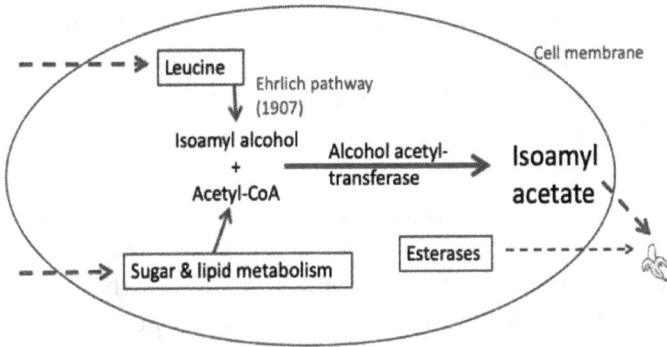

Figure 3.11. The major pathways involved in the production of fusel alcohols and esters from amino acid catabolism.

produced during fermentation, their effects on flavor and aroma and the corresponding taste thresholds in a typical beer matrix.

At high concentrations the fusel alcohols impart undesirable off-flavors, yet are important as an intermediary in the production of esters from the catabolism of amino acids via the Ehrlich pathway. As an example, Figure 3.11 illustrates the production of isoamyl acetate from leucine. Note that once the esters have been excreted out of the cell into the beer, they can be degraded during beer maturation by the action of esterases, enzymes which are also produced by the yeast. In addition to the individual yeast strain, many fermentation parameters can affect ester production; the most important being the level of oxygen, UFAs, relative availability of sugar and FAN in the wort and fermentation temperature. The presence of both oxygen and UFAs in the wort reduces ester production by inhibiting the expression of acetyl-transferase genes and there is some evidence that addition of UFAs to the wort can reduce the need for wort oxygenation prior to fermentation. As the ratio of sugars to FAN increases, as occurs with high-gravity brewing (OG in excess of 1.075), the production of both fusel alcohols and esters can increase from the higher level of acetyl-CoA, resulting in beer with exceptionally fruity, solvent-like characters. Increased FAN concentration in the wort can also increase fusel alcohol production, depending on the type of amino acid, especially valine, leucine and isoleucine, will affect the aromatic profile of fusel alcohols and esters. Increasing fermentation temperature increases the rate of amino acid transport into the cell and therefore has a positive effect on both fusel alcohol production and ester formation, especially with lager yeast.

One of the most important off-flavors, especially in the production of lagers, is the vicinal-diketone 2,3 butanedione (diacetyl). This compound imparts a buttery flavor and aroma to the beer, even at concentrations below 0.15 mg/L. As shown in Figure 3.12, although diacetyl is produced from acetolactate as a byproduct of the biosynthesis of the amino acids leucine and valine, during a subsequent period of beer maturation, the yeast can also reduce diacetyl to acetoin, a compound with a much higher taste threshold (greater than 150 mg/L). This topic is also discussed in Chapter 4.

Generally, beer is best when fresh — consumed as soon as possible after packaging. As time advances, various chemical changes occur that result in the reduction of desirable taste and aroma and the formation of staling compounds, primarily volatile aldehydes that are responsible for the characteristic cardboard sensation. These aldehydes may be from the oxidation of wort lipids, such as linoleic and linolenic acid or other unsaturated fatty acids, which survived the fermentation. There are various mechanisms of oxidation, including

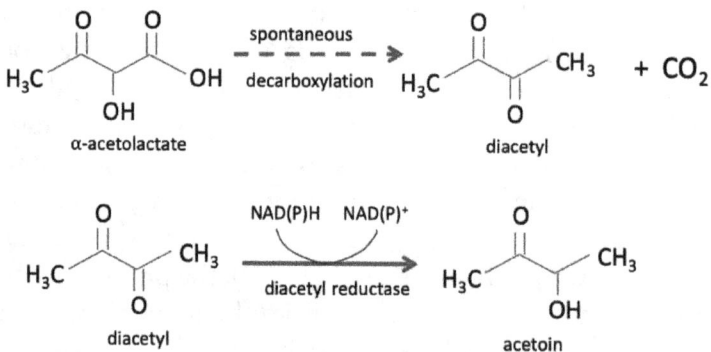

Figure 3.12.   Diacetyl formation from acetolactate and then reduction by the enzyme diacetyl reductase to acetoin.

enzymatic, autoxidation and photo-oxidation. Other compounds that may contribute to staling are Maillard products formed in the mash tun and kettle such as furfural and 5-hydroxymethylfurfural, or oxidation products from the hop acids. Many of these reactions resulting in staling aldehydes can be initiated prior to fermentation and packaging, depending on the quality of the raw materials and brewhouse operations that are employed. As much as possible, the prevention or slowing of these reactions is important for obtaining a good shelf-life for the beer as discussed further in Chapter 8.

3.6 Non-saccharomyces Species

Before the practice of using pure cultures of yeast was adopted in the late 19th and early 20th centuries, it would have been common to have multiple genera and species of microbes present in the fermentation. How well-adapted these microbes were to the brewing environment would determine their rate of survival and whether they would persist in sufficient numbers to affect the progress and outcome of the fermentation. An example of this brewing practice can still be found today in certain Belgian breweries producing traditional lambic-style beers by spontaneous fermentation. In addition to *Saccharomyces sp.*, lambic fermentations involve the activity of what are often considered as "spoilage" yeast and bacteria such as *Brettanomyces, Kloeckara, Pedioccocus* and enteric bacteria, which are naturally present in the brewing environment. These different microbial populations progress through a complex dynamic relationship spanning up to two years before completion and give traditional lambics their unique flavor and aroma.

Non-saccharomyces species of yeast can also be added purposely to create a special flavor profile. For example, *Brettanomyces bruxellensis* and *B. lambicus* have been used in co-fermentations with *S. cerevisiae*. Although *Brettanomyces sp.* are not very efficient at making ethanol and are slow growing under fermentation conditions, they do produce a lot of esters, especially ethyl acetate and ethyl lactate when in the presence of organic acids and ethanol. However, they can also produce acetic acid resulting in a sour character in the beer, and the objectionable "horse blanket" flavor originating from tetrahydropyridines.

A rather unique yeast that is new to the brewing environment is *Lachancea thermotolerans*. Originally used in co-fermentations of wine because of its ability to both acidify and produce large quantities of fruity esters, several wild strains have now been domesticated for use in the production of pure-culture sour beer. Although there is considerable phenotypic variability in this species, some strains have the ability to ferment maltose to ethanol at warm ale temperatures with rates and efficiencies comparable to *S. cerevisiae*. During the fermentation, these strains also produce lactic acid, depending on conditions up to 9 g/L, resulting in a final pH in the beer of between 3.2 and 3.5. This is comparable to other sour beers which utilize *Lactobacillus* bacteria to obtain the acidification but with much improved process simplicity and consistency from batch to batch.

3.7 Topics for Discussion

- How did the domestication of brewing yeast differ from methods employed with animals?
- What advantages does the brewer have when they propagate their own yeast?
- What do you think happens to a yeast population when it is re-pitched too many times?
- What problems might be encountered in the fermentation when high-gravity wort is used?
- What would be the most effective approach to lowering the ester production during fermentation?
- What potential problems could be encountered when trying to domesticate a wild strain of yeast for brewing applications?

3.8 Further Reading

1. Baert JJ, De Clippeleer J, Hughes PS, De Cooman L, Aerts G. On the origin of free and bound staling aldehydes in beer. *J Agric Food Chem.* 2012;60:11449–11472. doi:10.1021/jf303670z

2. Baker E, Wang B, Bellora N, *et al.* The genome sequence of *Saccharomyces eubayanus* and the domestication of lager-brewing yeasts. *Mol Biol Evol.* 2006;32(11):2818. doi:10.1093/molbev/msv/168

3. Boulton C. *Encyclopaedia of Brewing.* Chichester, West Sussex: Wiley Blackwell; 2013.

4. Gallone B, Steensels B, Prahl T, Baele G, Maere S, Verstrepen KJ. Domestication and divergence of *Saccharomyces cerevisiae* beer yeasts. *Cell.* 2016;166:1397–1410. doi:10.1016/j.cell2016.08.020

5. Guinard JX. Lambic. In: Thomas V, ed. Brewers Publications; 1990.

6. Hazelwood LA, Daran JM, van Maris AJA, Pronk JT, Dickinson JR. The Ehrlich pathway for fusel alcohol production: a century of research on *Saccharomyces cerevisiae* metabolism. *Appl Environ Microbiol.* 2008;74(8): 2259–2266. doi:10.1128/AEM.02625-07

7. Layfield JB, Sheppard JD. What brewers should know about viability, vitality, and overall brewing fitness: a mini-review. *MBAA TQ.* 2015;52(3):132–140. doi:10.1094/TQ-52-3-0719-01

8. Layfield JB, Vann LR, Sheppard JD. A novel method of inducing and retaining cell cycle synchronization in cultures of *Saccharomyces cerevisiae.* *J Am Soc Brew Chem.* 2014;72(2):102–109. doi:10.1094/ASBCJ-2014-0324-02

9. Pires EJ, Teixeira JA, Branyik T, Vicente AA. Yeast: the soul of beer's aroma — a review of flavour-active esters and higher alcohols produced by the brewing yeast. *Appl Microbiol Biotechnol.* 2014;98:1937–1949. doi:10.1007/s00253-013-5470-0

10. Piskur J, Rozpedowska E, Polakova S, Merico A, Compagno C. How did *Saccharomyces* evolve to become a good brewer? *Trends Genet.* 2006;22(4):183–186. doi:10.1016/j.tig.2006.02.002

11. Saerens SMG, Delvaux F, Verstrepen KJ, Van Dijck P, Thevelein JM, Delvaux FR. Parameters affecting ethyl ester production by *Saccharomyces cerevisiae* during fermentation. *Appl Environ Microbiol.* 2008;74(2):454–461. doi:10.1128/AEM.01616-07

12. Stewart GG, Hill AE, Russell I. *125th Anniversary Review: Developments in Brewing and Distilling Yeast Strains.* Wiley Online Library; 2013. doi:10.1002/jib.104

13. *The Practical Brewer.* Master Brewers Association of the Americas; 1999.

Chapter 4

Beer Finishing

4.1 Introduction

At the completion of the primary fermentation, the majority of the fermentable sugars have been consumed and the concentration of alcohol has largely been finalized. But the beer is not yet ready for packaging, as a number of important steps are still required before the product can be considered as finished and suitable for consumption. As illustrated in Figure 4.1, these steps include a secondary fermentation, also referred to as maturation, conditioning or lagering, yeast removal, carbonation, packaging and optionally a pasteurization step depending on the type of package and distribution strategy to be employed.

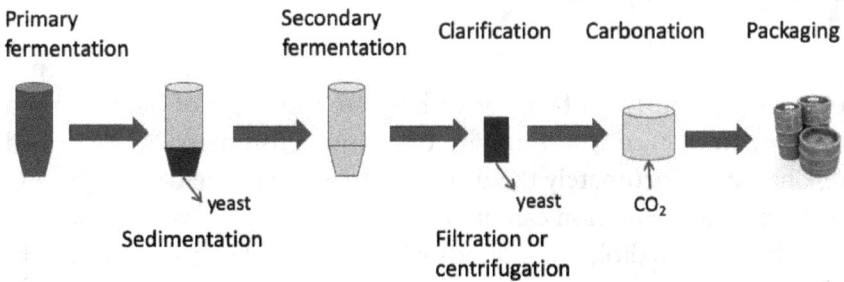

Figure 4.1.   Order of operations for beer finishing. Heat pasteurization may be employed following packaging to increase the shelf life of the beer.

4.2 Secondary Fermentation

The primary purpose of the secondary fermentation is to achieve a refinement of both the flavor and aroma of the beer by the continued metabolic action of the yeast and various spontaneous chemical reactions that occur. Other benefits include partial clarification by continued settling of the yeast and other suspended particulates, precipitation of possible chill haze as a result of the reduced temperature and an increase in the natural level of carbonation. The rate of progress in the secondary fermentation is based on several factors: the concentration of yeast in suspension, the availability of residual sugar to support its metabolism and the temperature. The first two factors are difficult to control, but predictability can be improved, if a measured quantity of fermentable sugar and yeast is added to the secondary fermenter, a practice referred to as *priming*. When the sugar and yeast is obtained as "green" beer from an active primary fermentation then the term *Kräusening* is used.

The maturation of bottom fermented beers is referred to as *lagering* and is typically performed at a reduced temperature (0–4°C) for an extended period of time (4–12 weeks), although this time is often reduced in North American practice. Effective lagering requires some residual sugar at the end of the primary fermentation (SG between 1.015 and 1.020) or the use of priming to maintain sufficient yeast activity to achieve the desired reduction in diacetyl and other undesirable flavor components. It is also important that the lagering vessel be closed and often pressurized at a low level to ensure the beer has no exposure to oxygen.

The quality of lager beers is particularly sensitive to the presence of diacetyl, a buttery off-flavor with a taste threshold of only 0.1 ppm or sometimes even lower among discerning customers. As described in Chapter 3, fortunately the diacetyl remaining in the beer at the end of primary fermentation can be re-metabolized by the yeast to acetoin and then butanediol, compounds with much higher taste thresholds. However, because lagering is usually performed at low temperatures, diacetyl reduction is a slow process. As shown in Figure 4.2, this can be circumvented by gently heating the primary fermenter to increase yeast activity for 18–48 hours prior to beginning the low temperature

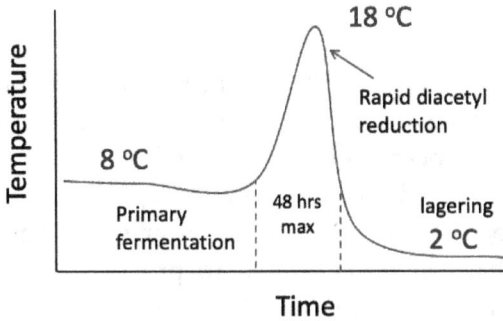

Figure 4.2. A typical temperature profile when a diacetyl rest period is employed.

lagering phase. Indeed, some breweries may even eliminate low temperature maturation, especially if a significant level of sugar adjuncts have been used to replace the barley malt.

Other significant changes can occur during maturation that impact beer quality, including transformation of hydrogen sulfide to sulfate, conversion of residual acetaldehyde to ethanol, a reduction in the harsh bitterness that may have resulted from high hopping rates and the potential for off-flavors being released from the autolysis of residual yeast, although this can be minimized with the use of low temperature.

The maturation or conditioning of ales is less critical than with lagers because some diacetyl character is considered to be acceptable in many ales, especially those with strong malt or bitter flavors. The secondary fermentation of ales is thus primarily to help clarify and carbonate the beer in a closed vessel, cask or bottle. The temperature used is much higher than in lagering, between 10 and 20°C, and usually for shorter periods of time, although this can be highly variable especially if cask conditioning is used. This is a traditional method in which the ale is stored in wooden casks made from a specific type of wood, often oak, and which may have been previously used for storage of different alcoholic beverages such as whiskey or bourbon. In this case, the cask will impart special flavors to the beer, the consistency of which often varies from cask to cask.

4.3 Clarification

Traditionally, an important aspect of beer quality has been the clarity or lack of visible suspended solids or haze. More recently with the growth of the craft beer market and the infrequent use of clarification technology, haze has become a more acceptable attribute to many customers, depending on the specific beer style. There are several technologies that can be employed to provide a clarified product, including sedimentation with or without chemical aids, centrifugation and various forms of filtration.

Although not usually the most effective, the simplest method for clarification is simply to allow the particles suspended in the beer to settle out over time. The rate of settling will be dependent on the particle size and difference in density between the solid and liquid phases. During primary fermentation, much of the yeast is in free-suspension as single or chains of cells, buoyed by the production of carbon dioxide gas. Strains of ale yeast have a tendency to attach to bubbles and float to the surface forming a mat, hence the name "top fermenting" yeast. In contrast, the cells of most lager strains will gradually sink during the primary fermentation and collect in the cone of the fermenter, hence the name "bottom fermenting" yeast, although in both cases when fermentation is active, there are many cells in suspension, especially the smaller younger cells. As the end of the primary fermentation is approached based on the availability of fermentable sugar, the cells will begin to flocculate or stick together in small clumps. Combined with the reduction in CO_2 production, the rate of sedimentation will increase significantly. However, it is not desirable to separate all yeast cells at the end of the primary fermentation, as some cells are still needed in suspension to achieve effective lagering or conditioning. Yeast settling will continue in the maturation vessel and so, by the end of maturation, the beer will be at least partially clarified. The level of clarity can be further improved by reducing the final temperature in the maturation vessel to 4°C or lower, a practice referred to as "cold crashing."

Yeast flocculation is the key factor in achieving efficient sedimentation. The mechanism of adhesion between yeast cells relies on the presence of specific proteins on the cell surface (*flocculins*) that are able to bind with mannose residues in the cell wall of other cells to which they come into contact. In this way hundreds or thousands of cells form aggregates that are visible to the eye and more readily settle from the beer. Several environmental factors affect the flocculation process, most importantly the presence of calcium ions, which are directly involved in the binding of the flocculins with carbohydrates on the cell surface. The presence of carbohydrates in the wort inhibits adhesion between cells, thus flocculation is usually delayed until the primary fermentation is largely completed. There are numerous genes in both ale and lager brewing strains that encode for flocculin proteins. Loss of flocculating behavior is apparently very strain-dependent but may be a result of either genetic changes in the culture (for example, those occurring after multiple re-pitching) or as a result of environmental factors associated with yeast handling procedures that can affect cell surface hydrophobicity among other things. Important environmental factors can include the degree of oxygenation of the culture prior to pitching or degree of saturation of the wort, temperature, pH, ethanol concentration and level of residual sugars. Refer to Table 4.1 for a summary of effects. The importance of environmental factors can also be dependent on the distribution of generational cell ages within the culture as affected by the method employed for cropping between serial re-pitches, with older cells exhibiting superior flocculation ability compared to virgin cells.

There are several different materials, referred to as fining agents, which may be added to the maturation vessel to aid in the rate of sedimentation. These include isinglas (made from the bladders of sturgeons), seaweed, Irish moss, gelatin and chitosan. In general, these materials function based on their polymeric properties and ability to neutralize the surface charge on the cells and other suspended particles, thereby encouraging coagulation. The solids that are removed should not be used as a source of yeast for re-pitching and should be disposed of. Even with the use of fining agents, relying on sedimentation to produce a clear, bright beer can produce inconsistent results.

Table 4.1.   A summary of the factors that can affect flocculation behavior within a yeast population. Note that all effects are strain dependent.

Factor	Effects on flocculation behavior
Oxygen availability	Aerobic propagation may decrease flocculation, wort aeration has a positive effect
Temperature	Generally no effect between 5 and 25°C
pH	Optimum range between 2.5 and 5 for most strains, high pH inhibits flocculation
Ethanol concentration	Generally a positive effect up to 10% ABV
Nutrient availability	Most residual sugars act as inhibitors, calcium ions are essential, magnesium and zinc may be beneficial
Cell age	Older cells are typically larger and more hydrophobic, virgin cells are smooth and also lack flocculins

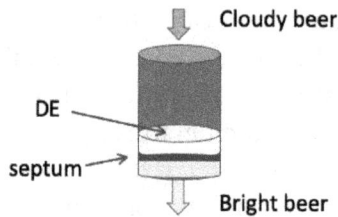

Figure 4.3.   An in-line filter packed with diatomaceous earth (DE) held in place by a perforated septum.

The most common alternative is to implement an in-line filtration process while the beer is being transferred from the maturation vessel to the bright tank. Until recently, the filtration medium of choice was *Kieselguhr*, microscopic particles composed of diatoms that have been ground and sterilized to produced *diatomaceous earth* (DE). As illustrated in Figure 4.3, the DE forms a filtration bed, supported by a perforated septum or screen through which the beer is pumped. As the filtration progresses, the depth of the bed is increased by periodic additions of more DE. After the filtration is completed, the DE and captured solids are disposed off. As an alternative to DE, filters may also use disposable filter cartridges made from cellulose or reusable

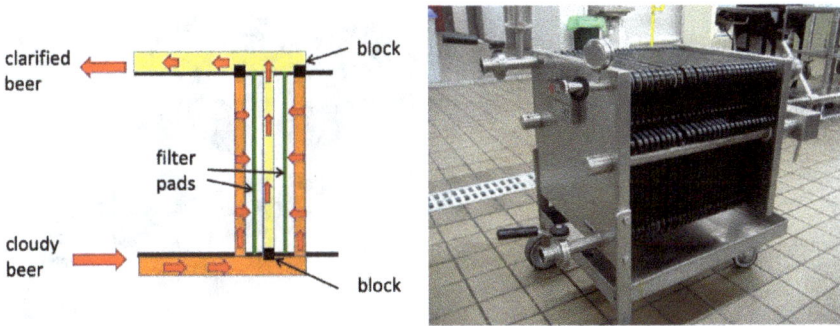

Figure 4.4. A plate and frame filter utilizing cellulose sheets sandwiched between multiple plates to provide a high surface area for removal of yeast and other suspended particles from the beer as it is pumped to the bright tank.

synthetic media. Cellulose pads or sheets are also used with a plate and frame filter apparatus as illustrated in Figure 4.4. The disposable sheets (40 cm by 40 cm or larger) can be of various pore sizes, fine enough to capture yeast and most other solids that can affect the visual clarity of the beer, but not fine enough to remove bacterial contaminants or haze caused by undissolved proteins. Multiple filter pads or sheets are configured in parallel, sandwiched between alternating plates that are blocked for flow at opposite ends so that yeast cake accumulates on one side of the filter pads. It is important to carefully control the pumping rate and associated pressure drop across the pads so that the cake does not become too compressed, thereby hindering the rate of filtration.

When operated correctly, filtration can be highly effective as a method for clarification. However, it is rather labor intensive, has relatively high operating costs and is time consuming, especially for clarifying large volumes. It also does not work as well with less flocculent strains of yeast, more commonly encountered when brewing at high gravity. To overcome these limitations, some breweries are now relying on continuous-flow centrifugation. In the disk stack centrifuge, as the beer is pumped to the inlet port, multiple perforated disks are rotated at high velocity to force suspended solids to the periphery of

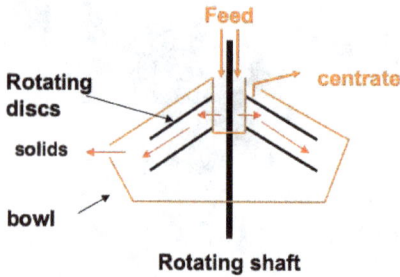

Figure 4.5.   The disk stack centrifuge uses multiple perforated plates rotating at high velocity to separate the inlet stream of beer into a continuous discharge of a concentrated yeast slurry and clarified beer.

the bowl where they are discharged as a concentrated slurry. The clarified beer, or centrate, leaves as a separate stream from the center of the bowl as illustrated in Figure 4.5.

Use of centrifugation is gaining in popularity based on its rapid and efficient continuous processing of large quantities of beer with minimal losses and low operating costs. There are, however, potential undesirable impacts on beer flavor and haze as a result of heat generation and rupturing of some yeast cells due to excessive shear forces. These effects can be minimized with proper controls on temperature and centrifugation speed, as yeast are generally very shear resistant.

4.4 Carbonation

Control over the level of dissolved carbon dioxide gas is essential to the quality of the final product, regardless of the beer style. The carbon dioxide is responsible for the production of foam or *head* on the beer when it is poured, the associated release of aromatics into the nose while drinking and the overall mouthfeel as the gas bubbles are released. However, too much CO_2 can also be problematic, resulting in beer gushing from cans and bottles when first opened and difficulties in pouring from tapped kegs without producing excessive amounts of foam. Although there is a certain level of carbonation that occurs naturally from yeast activity during both primary

Figure 4.6.   A 40 BBL stainless steel jacketed bright tank.

fermentation and maturation, generally final carbonation is accomplished in the bright tank (Figure 4.6) under controlled conditions of pressure and temperature.

The solubility of all gases in aqueous liquids is a function of both temperature and pressure. At a given pressure as the temperature of the liquid is decreased, the gas solubility increases, so, as beer is warmed, the carbon dioxide will begin to come out of solution resulting in a "flat" beer. Charts are readily available that contains data on CO_2 solubility over a range of both temperatures and pressures (for example, from the American Society of Brewing Chemists). At the end of fermentation without pressurization, depending on temperature, beer will typically contain between 1.2 and 1.7 volumes of CO_2 per volume of beer. This corresponds to between 0.6 and 0.85 pounds of CO_2 per BBL of beer at 0°C. Except for certain traditional ales, this level of natural carbonation is not sufficient for packaged beer and is increased to almost double this value while under pressure in the bright tank. The solubility of a gas is predicted by Henry's Law, which relates the effects of pressure on solubility through a coefficient specific for each gas, referred to as Henry's constant. For example, pressurizing to saturation at 10–13 psig would result in 2.5–2.8 volumes of dissolved CO_2 gas per volume of beer. Depending on style, somewhat lower or higher pressures can be used to produce the desired result. The time needed for saturation to occur depends on the dimensions of the bright tank and the means of introducing the CO_2 bubbles into the beer. The most common device used for this purpose is a porous stainless steel tube inserted into the bright tank through a sanitary port (often referred to as a diffuser stone as pictured in Figure 4.7). Pressurized CO_2 gas from either a cylinder or storage tank is reduced in pressure and attached to the quick-fit connection on the carbonation device. Fine bubbles of CO_2 are in this way introduced into the beer, while pressurizing the sealed bright tank. Depending on the size and dimensions of the tank, pressure is maintained for typically 2 or 3 days in order to achieve full saturation at a temperature between 0 and 4°C.

In addition to the use of a pressurized bright tank, with or without a diffuser stone, it is possible to carbonate more rapidly with an in-line diffuser or CO_2 injector as the beer is being pumped through a pipe between vessels or to the packaging operation. In some breweries beer has also been carbonated to saturation by introducing the beer into a previously pressurized vessel as a fine spray. Finally, although not common at a commercial scale, in-bottle carbonation is possible if there is some active yeast remaining in the beer. However, care must be taken to add the appropriate amount of priming sugar, otherwise there is

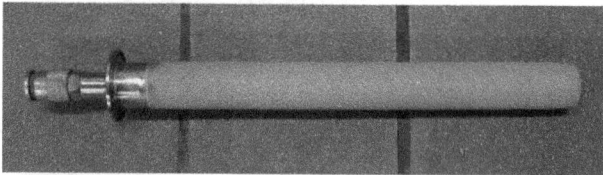

Figure 4.7.   A porous stainless steel diffuser "stone" for introducing bubbles of CO_2 into the bright beer. Note the quick-fit gas connector and sanitary flange for attachment to the bright tank.

the danger of developing excessive bottle pressure, even causing caps to pop (!), especially if the beer is not kept refrigerated at all times.

4.5 Packaging

Packaging of beer serves many different purposes: protects the beer from light and oxygen exposure, facilitates storage and distribution, and for customers, allows for brand recognition and a convenient way to purchase a specific quantity of their favorite product. Each different package option, whether kegs, bottles or cans, offers distinct advantages and disadvantages to the brewer, the customer and to society based on its potential environmental impact.

Stainless steel kegs are the universal package for beer sold as "draft," dispensed by means of a keg coupler and an external tap, usually situated remotely at the serving location, while the keg remains pressurized by an external source of CO_2 gas. Referring to Figure 4.8, kegs used in the USA are supplied in three sizes: 5.2 USG (approximately 1/6 BBL and referred to as a sixtel), 7.75 USG (1/4 BBL) and 15.5 USG (1/2 BBL). European and Canadian keg sizes are 20 L and 50 L. All stainless steel kegs are reusable and are typically identified and often tracked by each brewery to ensure they are returned by the customer or distributor, although a keg can remain in circulation

Figure 4.8.   The three standard US keg sizes: 5.2 USG, 7.75 USG (with reduced height) and 15.5 USG. The keg coupler attaches to the top "Sankey" connection.

for extended periods of time, so the brewery must have sufficient inventory of spare kegs to prevent a packaging bottleneck. Single-use plastic kegs are also available in some areas, however, these have not yet achieved wide acceptance due to cost and have also raised some environmental concerns. In several ways a steel keg is an ideal package for retaining beer quality, especially for non-pasteurized beer. The keg is pressurized with CO_2 to eliminate any ingress of air and because there is no light exposure, beer oxidation during storage is minimized. When a keg is returned to the brewery it undergoes a rigorous cleaning with a caustic solution and sanitizer, typically followed by steam sterilization and then pressurization with CO_2 gas before refilling. Kegs are not, however, very consumer friendly as they need to be kept refrigerated at all times, require a tapping system (see Figure 4.9) with a source of pressurized CO_2, while the dispensing lines need to be cleaned on a regular basis to prevent off-flavors from developing.

Bottled beer has now been available to the consumer for over 100 years and to many is still considered to be the ideal package, however, consumer preference is changing. Referring to Figure 4.11, bottles come in many shapes, sizes and materials, although glass is still predominant. There is a certain amount of standardization with regards to volume, the US bottles based on fluid ounces and the European based on milliliters.

Although bottled beer continues to be popular with both large and smaller craft breweries, there are several issues that are not in their favor. Perhaps the most significant problem is the potential for

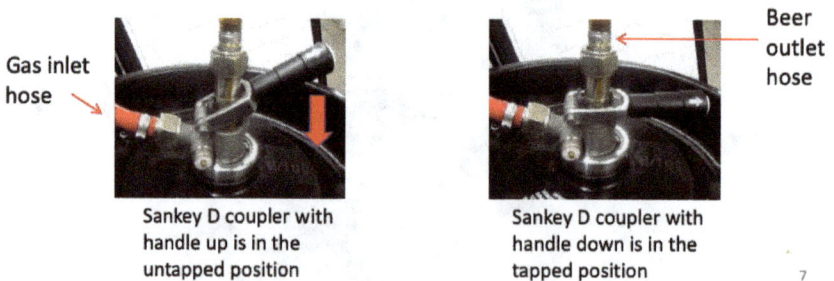

Gas inlet hose

Beer outlet hose

Sankey D coupler with handle up is in the untapped position

Sankey D coupler with handle down is in the tapped position

Figure 4.9.   The Sankey coupler is used for tapping commercial kegs. Low pressure CO_2 is connected to provide the back pressure to push the beer to the tap location.

Figure 4.10. An automated keg washing and filling system complete with pumps and recirculation tanks for caustic detergent and acidic sanitizer solutions.

12 oz.
standard
US long
neck

500 mL
standard
European

32 oz.
Howler

64 oz.
Growler

Figure 4.11. Common bottle sizes. Howlers and Growlers are typically filled in the tap room of a craft brewery for home consumption.

light-induced spoilage. This is caused by ingress of visible light through exposed bottles resulting in the conversion of one of the hop bittering acids, iso-humulone, to 3-methyl 2-butene thiol. This mercaptan gives the beer a distinctive and objectionable "skunky" odor. It can be largely solved if the bottled beer is protected from light exposure by the use of amber bottles stored in cardboard cartons or, alternatively if the bittering a-acids are enzymatically treated to remove the acyloin group. The second issue concerns oxygen exposure due to air entrapment during filling or ingress of air via the cap seal, although a well-run modern filling machine using high quality bottles and caps can largely eliminate these problems.

The environmental impact of bottle usage varies depending on local policies. In some situations, a deposit is charged at time of purchase to encourage bottle return to the retail outlet. The brewery must then collect the bottles and employ a bottle washing/sterilizing operation prior to refilling. In this way the bottles can be re-used several times, however, the energy expended in collection and washing may not always be justified and can constitute a market barrier for small craft breweries. An alternative is to package in new bottles each time, with the hope that the consumer will have the ability to recycle the bottles as part of a municipal waste collection process. In some markets recyclable plastic bottles made from polyethylene terephthalate (PET) are also used for beer packaging. Although PET bottles are often associated with lower quality products, the problems of light and oxygen spoilage have been minimized with layered construction and the use of amber color. Yet some consumers will maintain that the taste of the beer is affected by exposure to the plastic, especially if long term, thus limiting their widespread use.

The use of aluminum cans as an alternative to glass bottles is steadily growing. Overall, as of 2017 approximately 60% of the total beer sold in the United States was packaged in cans (https://www.beerinstitute.org/trends-beer-packaging/) and in 2019, the total volume of craft beer packaged in cans grew 17% to reach that of bottled beer, which declined 12.9% (Nielsen industry data from the Brewers Association). There are several reasons for the rising popularity of cans, such as providing a more effective barrier to exposure to both oxygen and light, lighter weight for reduced transportation costs and, in some

Table 4.2. A relative comparison of the three principal packaging options for beer. The number of ♦ indicates the strength of this factor for the specific package type.

Criteria	SS Kegs	Cans	Bottles
Product quality	♦♦♦♦	♦♦♦	♦♦
Environmental impact	♦♦♦♦	♦♦♦	♦♦
Consumer preference	♦	♦♦♦	♦♦♦♦
Cost to brewer per BBL	♦♦♦♦	♦♦	♦♦
Overall ranking	1	2	3

situations, more readily recyclable than glass. Generally, the use of cans has less environmental impact than bottles, however, the total cost to the brewer may be higher depending on the prevailing market conditions for aluminum and the quantities of cans purchased. Like bottles, cans are available in several different sizes and volumes, with the standard US can being 12 ounces and the European can being 500 mL. There continues to be a debate as to whether cans affect flavor, imparting a metallic character to beer, although the beer is not actually exposed to the aluminum. The inside of an aluminum can is lined with a very thin polymeric, epoxy-based coating to prevent corrosion by the beer. The composition of the coating is a proprietary formula developed by the can manufacturer and in the US, cans may contain the chemical bis-phenol-A (BPA) as a plasticizer. Unlike Europe and Canada, the US does not currently regulate the use of BPA.

4.6 Pasteurization

Named after the French scientist Louis Pasteur, the process of heat pasteurization is used for a variety of food and beverage products to reduce the level of microbial contamination and extend the shelf-life of the product. Depending on the characteristics of the product, type and quantity of bacterial contaminants, the temperature and time employed to achieve the desired effect will vary. An excessively harsh pasteurization process can result in adverse effects on product quality, while conditions that are too mild may not achieve the desired reduction in bacterial contaminants. Because beer is characterized by relatively high levels of alcohol and low pH, it provides an unfavorable environment for the survival of most bacterial species so effective pasteurization can be achieved at relatively low temperatures, typically 55–60°C, with heating times between 15 and 20 minutes or less. Large breweries will often employ continuous tunnel pasteurizers, in which canned or bottled beer will be subjected to a spray of hot water as it moves through the tunnel at a controlled rate. Alternatively, the beer can be pasteurized prior to packaging by passing it through a multistage heat exchanger. In this case, shorter contact times in the order of 15–30 seconds are used at elevated temperatures, between 71 and 78°C. This flash pasteurization process has the advantage that it can be also used for draft beer stored in kegs. Heat pasteurization is a relatively complex and expensive process for most small craft breweries to employ and can have adverse effects on taste if not precisely controlled. Thus, it may not be warranted, especially if the canned or bottled product is only sold locally with a rapid turnover. A less complex alternative for small scale operations would be cold filtration, in which an absolute filter cartridge with pore size of 0.45 micron or smaller is used as a final filtration step prior to packaging (see Figure 4.12). Since the beer is

Figure 4.12. A flow-through cartridge filter housing with sanitary connections.

not subject to heating there is less effect on beer quality due to staling reactions, although, there still may be some subtle changes in color and flavor. Depending on the material, some filter media can be cleaned and re-sanitized for reuse. Single-use disposable cartridges are typically more reliable, but also more expensive.

4.7 Topics for Discussion

- What are the advantages and disadvantages of using separate vessels for primary and secondary fermentation?
- Why is the presence of diacetyl more acceptable in ales than lagers?
- How could the procedure for cropping yeast from the primary fermenter affect the characteristics and performance of the re-pitch?
- How would yeast flocculation affect the performance of a disk stack centrifuge?
- How does the environmental impact of glass bottles compare to the use aluminum cans?
- Why is it important to have beer well clarified prior to pasteurization?

4.8 Further Reading

1. Boulton C. *Encyclopaedia of Brewing*. Chichester, West Sussex: Wiley Blackwell; 2013.
2. *The Practical Brewer*. Wauwatosa, WI: Master Brewers Association of the Americas; 1999.
3. Uhrich M. *Trends in Beer Packaging*. Beer Institute; 2019. https://www.beerinstitute.org/trends-beer-packaging/
4. van Bergen B, Sheppard JD. The effect of acid-washing and fermentation gravity on mechanical shear resistance in brewing yeast. *J Am Soc Brewing Chem*. 2004;62(2):87–91.
5. Vidgren V, Londesborough J. 125th anniversary review: yeast flocculation and sedimentation in brewing. *J Inst Brew*. 2011;117(4):475–487.
6. Watson B. *Bottles and Cans: Craft Beer Packaging Trends in 2018*. Brewers Association; 2019. www.brewersassociation.org

Chapter 5

Malting

5.1 No Barley — No Beer

Malting, and eventually brewing, started in the Middle East in a region that was referred to at the time as the Fertile Crescent, stretching from the Mediterranean Sea to Mesopotamia. There, dating back to 8,000 BC, as nomadic tribes transformed into agrarian societies, barley became one of the main crops, and allowed for two harvests a year. Our best understanding is that, the grain that had been left outside, at first by accident, sprouted in the rain. The sprouted grain then started naturally fermenting into a nutritious and intoxicating beverage. From there developed the most basic forms of malting and brewing.

For 10,000 years, barley has been the main grain used for malting and brewing. What makes barley uniquely fit for brewing are the following traits:

- Barley is a well-adapted small grain and grows in a wide range of soils and climates
- Strong husk, which protects the grain and make a natural filter bed for lautering process
- Malting traits are favorable for germination which leads to good modification of raw barley kernel structures
- Malted barley has a clean pleasant wholesome flavor

- Malted barley is the yeast's source for fermentable sugar, protein and trace elements
- Malted barley delivers the most comprehensive complex of enzymes out of all small grains

5.2 Why Do We Malt?

Barley grains are seeds, adapted to eventually grow into new plants. The kernel is designed to safely store the energy it needs to grow until growing conditions are conducive to germination. While the seeds are dormant, the nutrition within is stored as complex starch, a cell structure that makes the seeds hard as a rock, durable over a long time, and resistant to mold. Complex starches are not fermentable.

As illustrated in Figure 5.1, when a seed starts to grow, the enzymes stored in the germ of the seed activate, and convert the complex starches into simple sugars. This softens the seed, and makes the stored nutrition available to the growing plant. The job of a maltster is to convince the barley to germinate just enough to perform this transformation. Malting, like natural germination, (a) modifies the barley kernel to breakdown the cell walls in the starch compartment

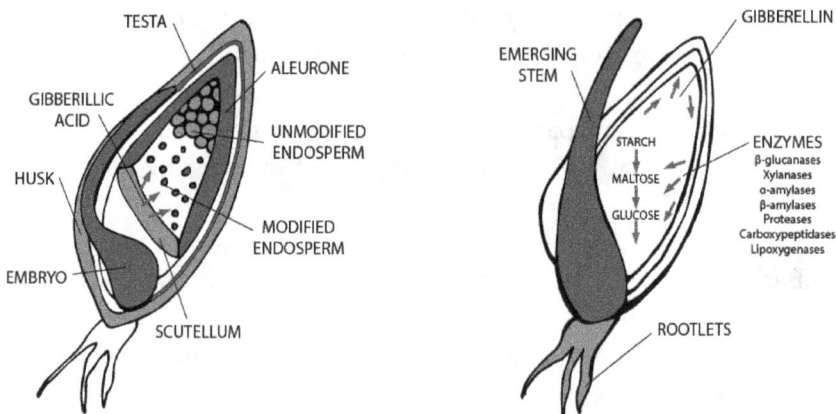

Figure 5.1. Schematic of the germination and modification process of barley through the activation of enzymes (used by permission of Epiphany Craft Malt LLC, 2019).

called endosperm, (b) activates the enzymes stored within the germ, which are necessary for the production of fermentable sugars and yeast nutrients. Finally, the maltster dries the germinated barley to suspend growth, and to develop color and flavor.

Fundamentally, malt provides the base for fermentation in beer by supplying fermentable sugars. The simple sugars in malt, when fermented, convert into alcohol. Beyond this basic function, nuances in the process of malting make it possible to brew hundreds of different styles of beers.

5.3 Creating Flavor Through Kilning

The wide variety of flavors and colors of malt are made by performing this same basic transformation in a carefully controlled environment. Color and flavor are developed through precise adjustments of heat, moisture, and timing, resulting in a vast catalogue of malts from a single type of grain. In addition to providing simple sugars, malts can be kilned or roasted to produce a variety of caramelized flavors. These flavors are created by chemical changes called Maillard reactions. This browning happens when specific sugars (aldose sugars) combine with protein (amino compounds), water and heat. This is a key flavor not just in malt, but also in caramel candy, bread crust and on the outside of a grilled steak. Resulting flavors can be nutty, toasty or bitter, depending on the specific conditions created.

5.4 Malting Small Grains

Malting, very simply, is a biological process transforming raw kernels of grain into a germinated form. Before we begin discussing the process, we need to start looking at the required inputs in more detail as it applies to small grains and their specific varieties.

Barley Types

As the key cereal for beer production, barley is planted and grown as a winter, spring or summer crop in two types. One type is called

SMALL GRAINS PLANT   HEAD   KERNEL

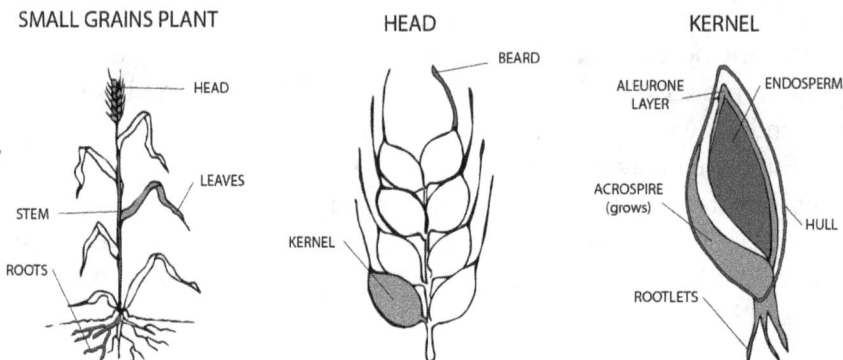

Figure 5.2. Schematic of small grains plant and barely kernel (used by permission of Epiphany Craft Malt LLC, 2019).

6-row barley as it grows with six kernels per row on the head, the other type is 2-row barley, which has only two kernels per row on the head. At present, the 2-row varieties are dominating the malting barley market as brewers prefer these plump kernels, which provide ample starch. The 6-row varieties are more tightly situated on the head, which keeps each grain smaller in size. This historically meant that 6-row barley had more enzyme activity by weight than 2-row. This used to be an advantage for brewers who were adding adjuncts as unmalted ingredients to their beers, as they needed the additional enzymes to convert unmalted sources of sugar. However more recently, breeding has made enough progress to the point where 2-row varieties have comparable enzymatic power and also enzymes have become commercially available as brewing additives. These changes have accelerated the shift towards 2-row.

Quality Requirements

The maltster has very specific requirements when receiving raw malting barley. To make the malting process homogenous, have a good modification and develop enough enzyme groups, to be acceptable, the delivered barley will need to fit the parameters shown in Table 5.1.

During the evaluation of barley quality, the maltster uses a representative sample to establish each benchmark. Moisture levels above

Table 5.1. Specifications as used by maltsters to evaluate the quality of barley before acceptance of a delivery.

Malting Barley Quality Selection Criteria	
Characteristic	**Specification**
Moisture	<13.5%
Protein	9.5–12.5% (dry basis)
Plumpness	>90% (6/64" screen)
Damaged or Broken	<2%
Varietal Purity	>98%
Germination Energy	>95%
DON	<1.0 ppm

13.5% pose a challenge to good storability. If the small grain moisture is higher, microbial growth and germination loss are elevated. Protein levels in barley and other grains vary based on growing conditions, agronomic practices and plant genetics. Higher protein levels (above 11.5%), limit extract potential. While, lower protein levels (below 10%) limit enzyme development. Plumpness can be determined by using screens that have different size slotted sieves: 7/64, 6/64 and 5/64 inch openings. Plump barley is associated with higher starch content and 2-row barley is considered plump at >80% over 6/64 inch sieve. As the maltster looks at the different screens, uniformity of size helps with an even water uptake during steeping and an even modification.

"Germination energy" measures the potential of kernels to actively germinate, once conditions are met. Functional malt production requires most, if not all, kernels to sprout at an even rate. This allows all of the staches to convert at the same time, maximizing yield and brewhouse performance.

The final important analytical test is the Deoxynivalenol (DON) value. It measures the mycotoxin produced by fusarium mold. This is important to brewers as this compound can make it all the way into finished beer where it leads to foaming over of carbonated beers due to the surface area it provides. This beer quality problem is called gushing.

Beyond the specifications that are measurable during the chemical evaluation, a maltster also performs a sensory evaluation, looking at the brightness and evenness of color, the smell of the grain, and scanning for damaged or broken kernels and signs of insects or insect damage. Additionally, the maltster looks for damage to the kernels from harvesting, grain handling, heat exposure, frost, transportation or storage that can lead to lower malting extraction, poor performance or off-flavors due to low germination, compromised vigor or contamination. When selecting barley for malting, good quality malt only comes from good quality barley.

Raw Grain Storage

Once the barley meets malting quality specifications, storage conditions and management become very important to maintain the highest possible quality for up to two years. During storage, temperature and moisture levels are crucial to maintain germination energy while keeping insects away, mold and fungal contamination low and to prevent premature germination. Some barley varieties are equipped with a natural dormancy that protect the grain from pre-germination on the stalk immediately after maturation (harvest time). A few weeks to months are sometimes needed in storage to break this dormancy, which would keep the grain from sprouting even when conditions are favorable to do so. When moisture levels are too high, drying down and further maturation of grains is possible to increase storability and keep germination high over time. The operator needs to exercise caution as high temperatures and over drying can destroy germination.

Chemical Composition

Comparing raw malting barley with finished pale malt, small subtle differences are shown in the Table 5.2. Of note to the reader is that both sets of values are "as is" % of mass fractions and include

Table 5.2. Comparison of raw versus malted barley.

Parameter	Malting Barley		Malt	
	As is	Dry matter	As is	Dry matter
Moisture	12–14		4–5	
Starch	55–57	64–66	56–58	58–60
Other extract	11–13	13–15	16–18	17–19
Protein	10–13	12–14	9.5–11.5	10–12
Fiber	4	4.5	5	5.2
Minerals	2.5	2.8	2.4	2.5
Fat	2	2.3	2	2.1

Source: Narziss/Back, Die Brauerei Band #1, 2012.

moisture, which in malt is less than half than what is found in raw barley. To ensure storage quality the moisture content of barley is critical and 12–15% is normal to do so successfully. The most important carbohydrates in malting barley are the alpha-glucans which make up amylose and amylopectin of the starch. The next group of carbohydrate components are beta-glucans which are cellulose, hemicellulose and gums. Other components are pentosans and trace amounts of sugar. Of course, the protein content is particularly important for maltablility, yeast nutrition, foam stability, taste and the colloidal stability of beer.

5.5 Other Malted and Unmalted Grains Used in Brewing

Many small grains, both malted and unmalted, can be used to supplement malted barley for brewing beer. All unmalted grain introduces unmodified starches and complex fibers that need to be addressed in the mashing phase to allow for a successful brew. When adding raw grain, it is important to manage the ratio of malted to unmalted grain, and to ensure that the unmalted starches are broken down by milling and/or cooking before mashing with active malted barley.

Wheat

When wheat is malted it develops a sufficient quantity of enzymes to be functional and allow for recipes of over 50% of the grain bill. The higher protein content leads to strong head retention in the finished beer and adds a pleasant mouthfeel. The extract contribution of wheat is as high as that of malted barley. However, a strong husk protecting the kernel is missing, which makes wheat harder to separate from the liquid mash in the lauter tun. The malted wheat makes for a dense filtration bed and increases the viscosity of the wort.

Rye

In a number of beer styles and distilled spirits the use of malted rye is primarily driven by the rye's distinct flavor contribution. The taste of rye is often described as bitter or spicy, where the aroma has a savory undertone. Rye beer in many cases tends to be pronounced bitter and dry.

The kernels of grain, or berries as they are referred to, are smaller and somewhat crooked.

Others

Many other small grains can and are malted to access their function and flavor, such as corn, spelt, buckwheat, rice, sunflower seed, millet, sorghum, emmer, peas and triticale to name a few.

5.6 How to Malt

Malting has only two ingredients, grain and water. This already hints at the importance of understanding and controlling the process. On the one hand, the mere fact that the grain is sprouting does not automatically lead to a functional malt. On the other, being able to repeat the outcome batch after batch is a key to a successful malting operation. Figure 5.3 is an overview of the basic steps in the process.

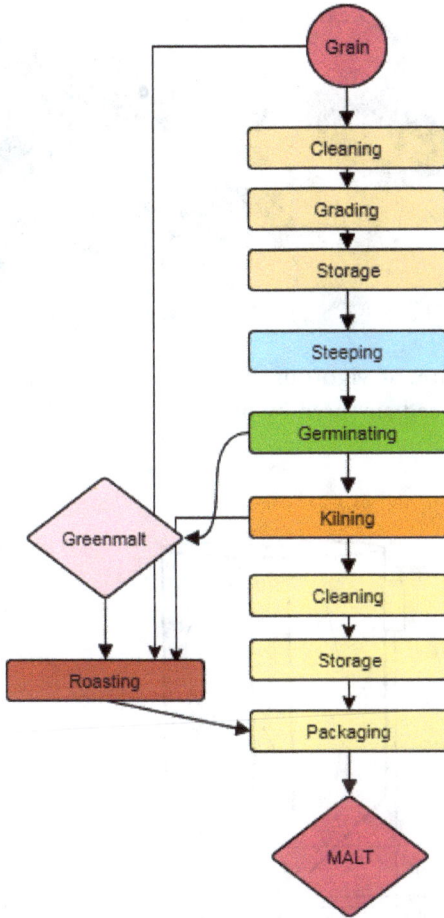

Figure 5.3. Flow-chart of the malting process (used by permission of Epiphany Craft Malt LLC, 2019).

Steeping

The malting process starts with rehydration of the grain from about 12% moisture during storage to over 42%. This first step toward waking up the kernels to begin germination is done in large cylindric-conical vessels or large round flat bottom steeps as illustrated in Figure 5.5. In either vessel, the grain is submerged in water, and

Figure 5.4. Picture of vigorous steeping in a cylindric-conical vessel by introduction of forced air (used by permission of Buhler Inc., 2019).

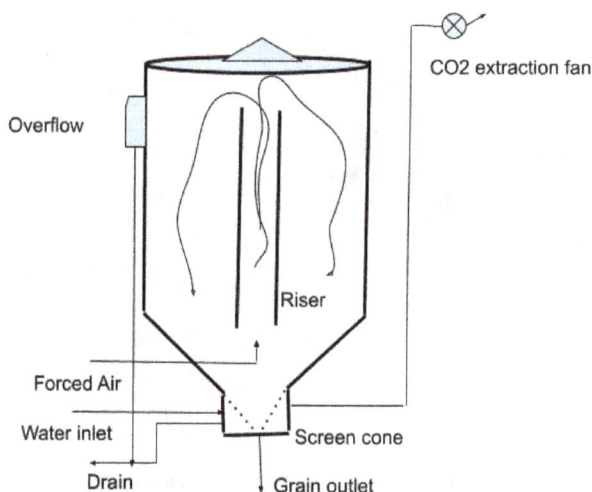

Figure 5.5. Schematic of a modern steep vessel allowing for mixing, aeration and dry rests based on the set-up shown. (used by permission of Epiphany Craft Malt LLC, 2019).

vigorous aeration is required (see Figure 5.4). The turbulence created by aeration helps clean the kernels with water movement and physical rubbing of the kernels against each other. After a few hours of steeping, bad and damaged kernels called "floaters" along with debris and dirt are washed out by overflowing the steep tank.

Signs of germination can start at moisture levels as low as 30%, although at least 42% is necessary for the creation of functional green malt. Besides moisture, the other key aspect during steeping is oxygenation. Barley and other small grains show sluggish germination when submerged, so it is important that air bubbles introduced into the water-grain mixture provide sufficient oxygen to avoid stunting growth. There are two distinct phases during steeping: the wet steep followed by a dry steep. Wet steeping is the active hydration with the barley steeping in cool fresh clean water. While hydrating the water-grain mixture is aerated to encourage chitting and agitated to wash dust and dirt from the grain. This is followed by a dry steep where the water is drained out via a straining port and the barley remains in the steep vessel with only a CO_2 removal fan pulling from the bottom to ensure fresh air supply. In summary, the goals are hydration, oxygenation and cleaning of the grain.

Germination

After the initial steeping to hydrate the grain, the germination phase is performed in a separate vessel as shown in Figures 5.6 and 5.7, and split into three distinct stages: chitting, sprouting and greenmalt.

During these three stages, the grain undergoes a metamorphosis on the inside with visible signs of rootlets and acrospire on the outside. Aided by moisture, temperature and the enzymes within each grain, starches convert into simple sugars. During germination the maltster's main focus is on temperature, moisture and airflow to arrive at a greenmalt with a definite composition without the development of a new plant.

Chitting

The first sign of germination is when you observe small white tips emerging on the end of the kernel where the Endosperm is located. These tips are called chitts, and those start to be visible within 24 hours of the first steeping. Now the kernel metabolism is fully active, converting the stored energy that would fuel the first stages of

Figure 5.6. Picture of the inside view of a round germination vessel with all core equipment components shown: slotted malting floor, turning machine with its spirals, round side walls with turning machine track (used by permission of Buhler Inc, 2019).

Figure 5.7. Picture of a malting floor from underneath showing the support structure and open space for forced air distribution (used by permission of Buhler Inc., 2019).

plant growth. The maltster's job is to ensure that the CO_2 and heat that are created by this process are efficiently removed by venting cool fresh air through the grain bed. Once chitting has occurred in over 90% of the seeds, it is time to adjust for the desired target moisture.

Sprouting

Referring to Figure 5.8, here we can observe actual rootlets growing out of the Endosperm. The other main plant part, the acrospire or leaflet, that is growing during this stage is harder to see as it is hidden under the husk. In order to see it, the maltster has to peel off the husk carefully. This is done to check on how far along the germination or modification of batch is. In the old days, a maltster knew that modification was complete when rootlets matched the length of the kernel and the acrospire was ¾ the length of the kernel under the husk. Depending on the parameters of the batch, this occurs 4–7 days after initial steeping. During this stage the grain bed in the germination vessel is cooled by flowing fresh temperate air through it. To further control the germination speed, the forced air coming in can be blended with recirculated used air returning from the germination box. This leads to a CO_2-rich air supply, which will lead to slower metabolism in the faster germinating kernels, thereby resulting in a more homogenous modification.

Figure 5.8. Picture of germinating barley kernels (used by permission of Epiphany Craft Malt LLC, 2019).

In today's malting toolkit, we use a modification testing device called the Friabilimeter. This rotating screen drum crumbles up 50 g of malt over the course of 8 min to give the maltster a very accurate number reflecting the degree of modification as a percentage. Depending on malt type friability of fully modified malt is between 78% and 92%.

Greenmalt

As high modification is reached, all enzyme groups are well developed and we reach the greenmalt stage. The maltster primarily worries about batch homogeneity and beta-glucan levels. In malting, homogeneity means that almost all kernels are at the same modification rate. Uneven modification would force the brewer to fight with a mash that carries a lot of over-modified and under-modified grains at the same time, which leads to lower yields and slower run-offs. Over-modification leads to lower yields as the growth of the rootlets and acrospire consumed more seed energy than necessary. Under modification or incomplete modification results in complex starches remaining in the kernel, which cannot be fermented, and is tied to high beta-glucan levels. Well-modified malts show a level of 40–80, at levels over 100 brewers will see a yield loss because the beta-glucans block sugar clusters from dissolving into the wort. As the levels of beta-glucan exceed 200, brewhouse lautering times increase and wort viscosity goes up, which leads to a further loss in yield.

Kilning

Once the grain is fully modified, the maltster stops the modification process by removing moisture, which stops all chemical and biological transformations. A combination germination-kiln vessel is illustrated in Figure 5.9. In the first phase, called withering, the grain is flooded with gentle dry air. This dries the kernels without shocking the enzymes or altering the color and flavor of the grain. After withering, we raise the temperature to force out additional moisture; this is called force drying. Finally, we adjust the temperature again to

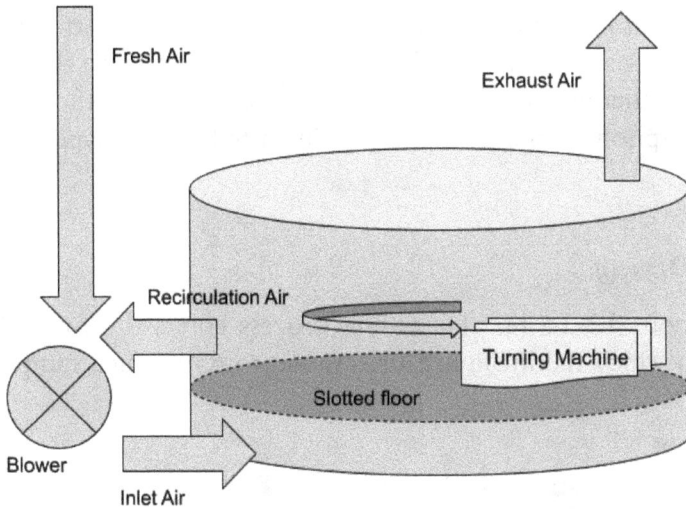

Figure 5.9. Schematic of a round Germination-Kiln-Vessel showing the main components with focus on the air-flow management (used by permission of Epiphany Craft Malt LLC, 2019).

develop color and flavor through Maillard reactions, in the phase called curing. At the end of germination, the greenmalt has about 38–48% moisture, and by the end of the kilning process the malt will have only 3–5%. By that the maltster has made a stable, storable product. The three phases combined take 24–48 hours to complete. The air blower capacity is at a minimum of 3500 m3/h.

Withering

The initial phase of drying, or withering, is also called free drying. The inlet air starts out at 50°C and raises up to 60°C, over the first 8–10 hrs. Throughout this time, the air exiting temperature of the kiln remains around 25–35°C, because the evaporating moisture cools the air in the malting box. Once the grain reaches 22%–18% moisture, the temperature of the exiting air quickly catches up to the inlet air temperature. This moment is called breakthrough. This effect is because all available moisture during the withering stage has been removed, taking away the evaporative cooling effect. It signals to the maltster that it is time to raise the kiln temperature to start force drying. When making smoke malt, smoke is introduced into the air during the withering stage. The high moisture content of the malt aids the absorption of the smoke flavor imparted by the type of wood used.

Force Drying

After about 12–14 hrs the kilning process enters the force drying phase in which the incoming air is raised to 65°C, then ramping up to 70°C.

Curing

After drying to below 8% moisture content is reached, the final stage of curing takes the malt to its final level of 3–5%, driving out any undesirable green-grassy vegetable-like flavors. Reaching a holding temperature of 82–84°C is critical to accomplish this goal.

5.7 Roasting

An important part of the malting process is to create color and flavor compounds that shape the character of a beer style. Roasting is very different from malting. In fact, most malting operations do not have a roasting facility. As pictured in Figure 5.10, drum roasters are operated to produce specialty malts that are generally less than 10% and often only 1–3% of a brewer's malt bill. Aside from the lower volume needs, roasting drums are inherently dangerous and can cause smolder fires due to the extreme heat conditions required to produce colors in grains and malts. They can reach temperatures of 240–260°C rather than a kilning floor that might reach 110°C. This much higher heat exposure to the grain or malt is possible because the steel drum that holds the product is directly above a burner that heats the drum from underneath, as well as quickly warming the air passing through and around the drum. To maintain an even heating of the grains the drum turns at about 30 rotations per minute. The kernels are evenly toasted by direct contact with the drum walls,

Figure 5.10.   A drum malt roaster with cooling table (used by permission of Epiphany Craft Malt LLC, 2019).

as well as by airborne kernels dropping through the air flow. To keep the heat and potential fires in check, roasting drums are equipped with a water spray pipe in the center of the drum. In Figure 5.11, the schematic shows the basic set up of a roaster, which has three main components: 1) burner with air ducts, 2) drum with air ducts, 3) cooling table. The cooling table is a screen on which the finished roast is unloaded. The spreader arms evenly distribute the roast across the table, while ambient cooler air is being pulled through hot product. The rapid cooling stops any further unwanted color and flavor development and it allows for quicker removal of the batch to a storage container or silo.

The roaster operator differentiates between two kinds of heating the product, called conductive and convective roasting.

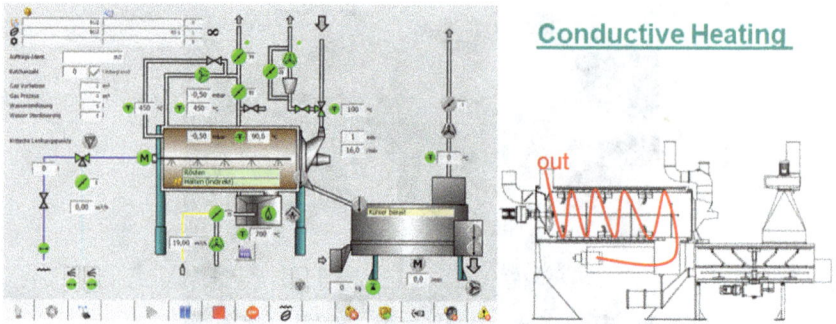

Figure 5.11.   Schematic of a drum malt roaster set to a conductive heating mode where only the outside of the drum is heated (used by permission of Buhler Barth GmbH, 2019).

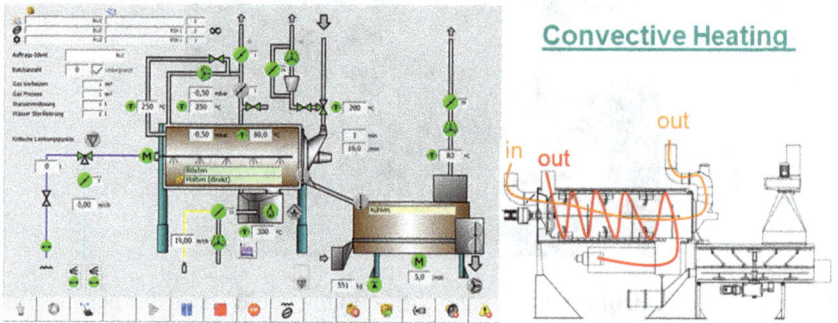

Figure 5.12.   Schematic of a drum malt roaster set to a convective heating mode where the outside of the drum is heated first then the hot air is directed to pass through the inside of the drum before exhausting (used by permission of Buhler Barth GmbH, 2019).

In the schematic the red line describes the conductive heating function, in which the burner heat flows around the outside of the drum rising up from the fire box underneath. Only the drum warms up and indirectly heats the malt on the inside.

As shown in Figure 5.12, the roaster is set to a convective heating function, in which the heated air passed through the center of the drum after first passing around it. The exposure of direct hot air to the product heats it more quickly and toasts, colors and dries at the same time. As we now look closer to how different inputs are roasted, it will become clear why this is important in the roasting of malt products. Generally roasting of grains and malts take between 90 min and 3 hours, followed by the cooling step once the target color is reached. The cooling step is very important.

Raw Grain

In order to create a very dark color in the roaster, the product needs to have a dense internal structure to conduct heat quickly and evenly. Therefore, we use raw barley to make the darkest color ingredients called Roasted Barley with a color range of about 500–580 SRM. To achieve this, the roaster uses the convection heating setting and applies heat continuously for the next 90 min until the drum reaches 440–450°C. Before the operator unloads, the roasted barley is quenched, or precooled, to slow the color development and avoid any potential smoldering of kernels before the product hits the cooling table.

Finished Malt

The roasting of finished dry malt creates color and flavor starting at about 40 SRM with biscuit type notes all the way to 350–450 SRM chocolate notes. As with the process for roasting raw grains, the process takes about 90 min or more and is accomplished using convection heating, driving color and flavor into the malt while resulting in even lower moisture content.

Green Malt

The key attribute of green malt is that it has not yet been kilned and, therefore, has retained all of its moisture. When roasting caramel or crystal type malts, the high moisture and enzymatic power still active in the green malt is essential to success. Crystal malts are roasted in a two-step process. The first step is to liquify the sugar created during the malting process by "mini-mashing" each kernel on the inside.

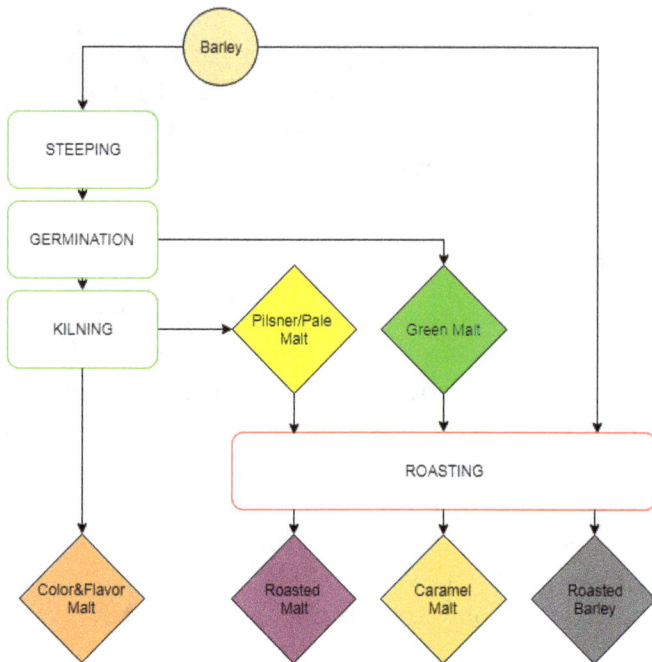

Figure 5.13.   Flowchart of the four main categories of malts produced and the main pathways of how a maltster makes choices during the malting process to determine character of the final malt product (used by permission of Epiphany Craft Malt LLC, 2019).

This is accomplished by conductive heating the kernels to about 70°C for one hour and retaining the moisture in the drum. Following this saccharification is the convective heating step that quickly removes the moisture while crystalizing the sugar inside the malt. Depending on the color and flavor intensity ranging from 10–180 SRM the second step can take up to two hours.

A good overview of the different roasting inputs and its relating outputs is shown in Figure 5.13.

5.8 Malt Storage

Malt storage is important at the malting operation, the wholesaler and also for the end user. Malt in 25 kg bags or malt silos that do not let moisture in will allow for storage up to 18 months and longer. Properly cleaned and finished malt show no signs of rootlets or damage to the husk at a moisture ranging from 3–5%. Any of these three parameters being off specification will result in inferior storability. Besides flavor loss, stored malt is at risk of damage from rodents, mold growth as it picks up moisture and insects. Temperature and moisture control in the storage facilities are the keys to effectively manage large inventories.

5.9 Topics for Discussion

- What ingredients are needed for malting?
- What does malt grain provide for Brewing and Distilling that raw grain does not?
- Which compounds in the grain are responsible for the functional changes during malting?
- How many days does the malting process take on average?
- What are the three main process steps in malting?
- Which are the parameters a maltster has control over?
- Why does plumpness play a key role for barley selection?
- How does the maltster create color and flavor?
- How long has humanity been malting?

5.10 Further Reading

1. *European Brewery Convention: Malting, Manual of Good Practice*; 1998.
2. Fachverlag HC, Joseph H, Ludwig N, Werner B, Die B. *Band #1: Die Technologie der Malzbereitung*. Wiley-Vch Verlag; 2012.
3. Images: Courtesy of Buhler Inc, Brewing & Malting, Plymouth, MN.

Chapter 6

Hops in Beer

6.1 Introduction

Hops (*Humulus lupulus* L.) are a species of the hemp family (Cannabaceae) and belong to the same order as nettles (Urticales). It is a dioecious plant i.e. there are female and male plants. However, only the female plants are able to develop hop strobilus, also known as hop cones. In the brewing process unfertilized hop cones are used. The pollen coming from the male plant can pollinate the female cone carried away by the wind. However, pollination is undesired as the seeds add weight to the overall product and have no brewing value. Therefore, male plants are only needed for the creation of new varieties through cross breeding and tend to be eradicated from the regions where female plants are cultivated [1].

6.2 Hop Cultivation

The hop plant is perennial. Every spring new shoots emerge from the rootstock, but only two to three are trained on wires or coir (coconut rope), depending on growing region. The rest are manually removed. The shoots allowed to grow have hooked hairs that allow them to climb around the supporting trellis in a clockwise direction. The bines can grow between 10 and 30 cm per day and are able to grow longer than 7 m (see Figure 6.1). The color of the bine varies among cultivars and

Figure 6.1. Hop field.

ranges from reddish purple through degrees of mottling to various shades of green [2].

In the northern hemisphere, the hop plant begins its growth in April. When the plant has reached a certain height, before reaching the end of the trellis, it develops lateral branches and after some time they begin to flower. The bloom of the hop cones takes place from the beginning to the middle of July due to decreasing day length. The dependence on the number of daylight hours (up to 18 h/day) relegates hop cultivation to regions between 35° and 55° latitude (see Figure 6.2). After flowering, cones develop rapidly and the hop plant is mature. Rainfall, especially during development of the umbels, is beneficial for the harvest and for the brewing quality [3].

Harvest takes place between the end of August and the middle of September depending on the variety and the weather. The hop plant

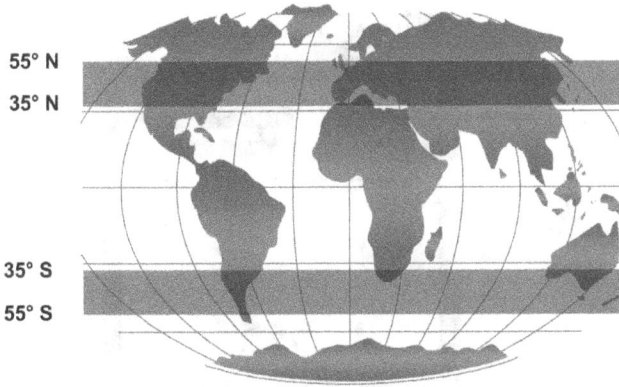

Figure 6.2.   Latitudes where hops grow.

is cut on the bottom by a hop-harvesting machine and pulled off from the top. The right time for harvest is choosen by the grower and compromises the yield, the brewing quality and the presence of diseases.

Hops are also grown in the Southern Hemisphere between latitudes 35° and 55°. Here, hop cultivation takes place between October and March.

After harvesting, the cones are separated from the rest of the plant with a machine called a picker. With mechanical harvest a weight loss of about 10% can be expected through the loss of cones and lupulin [3].

As soon as a hop plant is removed from the field, a degradation process begins. Therefore, drying must occur as soon as possible. After harvesting, the hop cones have a water content of about 80% and after drying they should reach a moisture content between 9 and 11%. Hot air kilns are available in hop cultivation farms for this purpose [3]. Figure 6.3 shows a cross section of a hop kiln. The drying process varies among varieties, the weather conditions and the quality of the cones:

- 4–6 hours kilning
- Temperature 62–65°C
- Air speed 0.3–0.4 m/s
- Height per layer: 30–35 cm

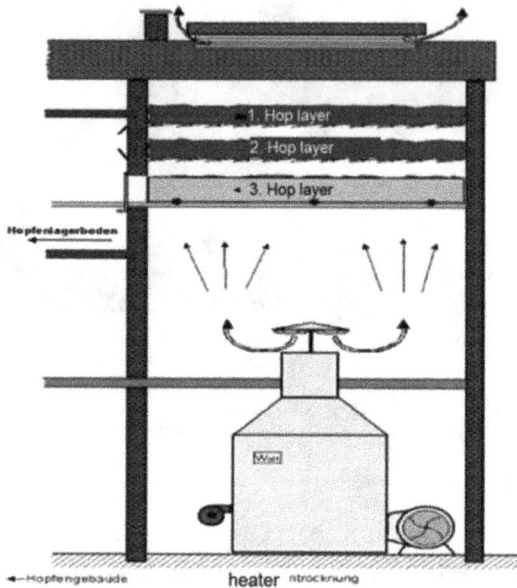

Figure 6.3.   Cross section of a hop kiln.

Source: LfL, Grünes Heft 2011.

Due to the cone structure, drying needs the experience of the grower and has to be carried out gently. The strig contains a larger water content than the bracts and bracteoles (see structure and composition of the hop cone for more information). If drying is done with excessive heat, the bracts and bracteoles will dry excessively whilst the strig will remain moist. If improper drying is done, the hop cones will spoil once baled.

Conditioning is carried out after drying, to allow the moisture differences between the strig and the bracts to adjust. The hops are conditioned in a temperature and moisture-controlled chamber at 20–24°C and a relative humidity of 58–65% [1]. After conditioning, the dried hop cones are packed into square bales.

6.3 Hop Growing Regions and Main Varieties Grown

About 97% of worldwide hops are destined for brewing purposes. The two main growing regions are the United States of America and

Table 6.1. Hop acreage of the top 10 countries and worldwide between the years 2008 and 2017.

Country	2008	2009	2010	2011	2012	2013	2014	2015	2016	2017
USA	16.551	16.077	12.662	12.055	12.923	14.254	15.707	18.478	21.570	22.576
Germany	18.695	18.472	18.386	18.228	17.128	16.849	17.308	17.855	18.598	19.543
Czech Republic	5.335	5.307	5.210	4.632	4.366	4.319	4.460	4.622	4.775	4.945
China	5.683	6.023	5.502	4.458	3.989	2.831	2.655	2.320	2.639	1.648
Slovenia	1.577	1.579	1.391	1.379	1.160	1.166	1.296	1.406	1.484	1.591
Poland	2.233	2.167	1.867	1.564	1.510	1.407	1.410	1.444	1.475	1.576
England	1.071	1.081	1.070	1.114	1.054	985	929	895	920	967
Australia	484	514	448	454	452	449	408	488	546	631
Spain	465	469	508	533	541	485	535	543	540	521
France	801	533	580	500	439	381	431	440	459	481
World-wide	57,297	56,747	52,029	48,529	46,971	46,246	48,172	51,512	56,141	58,739

Source: The Barth-Report [4]

Germany. According to the Barth-Report 2017–2018 [4], these two countries account in 2017 for more than 75% of the total hop production and more than 70% of the total acreage. Other hop growing countries are the Czech Republic, China, Slovenia, Poland, England, Australia, Spain, France, New Zealand and South Africa (see Table 6.1).

The main growing region in the United States of America is located in the Pacific Northwest region (Washington, Oregon and Idaho States). The United States total acreage outside the PNW is slightly less than 4.5%. The industry-leading variety in the United States of America is Cascade followed by Centennial, Citra®, CTZ and Simcoe® accounting for more than 50% of the acreage. The average farm size continued its slight upward trend to 312 ha for crop 2017 compared with an average of 307 ha for the previous year [4]. Within the PNW region, the average hop acreage cultivated by Washington growers was 432 ha, followed by Idaho and Oregon with 283 ha and 138 ha, respectively [4].

In Germany, there are four major growing regions, Hallertau, Elbe-Saale, Tettnang and Spalt; Hallertau being by far the largest. The main varieties grown in Germany in 2017 were Herkules, Perle, Hallertau Tradition, Hallertau Magnum and Hersbruck Spaet. These five varieties account for more than 73% of the acreage cultivated. The farm structure in Germany is characterized by small farms compared to those in the United States of America. There are 912 hop producers with an average acreage farm size of 17.9 ha [4].

6.4 Structure of the Hop Cone

The hop cone consists of a strig, bracteoles and bracts and the lupulin gland where a yellow powder called lupulin is present (see Figure 6.4). Lupulin glands are more abundant on bracteoles than on bracts [2]. It is in these lupulin glands that the main brewing principles of hops, the resins, the essential oils and the polyphenols are synthesized and accumulated [5, 6]. Hops are unique and not like other plants in that they possess lupulin glands, which contain hop acids, resins and essential oils. This unique resinous powder is assumed to serve as protection for the hop plant against birds and insects and also against microbiological attack due to its bitterness.

6.5 Composition of the Cone

The typical average composition of the dried raw hop cone is shown in Table 6.2. The total resins, also known as bitter compounds, the

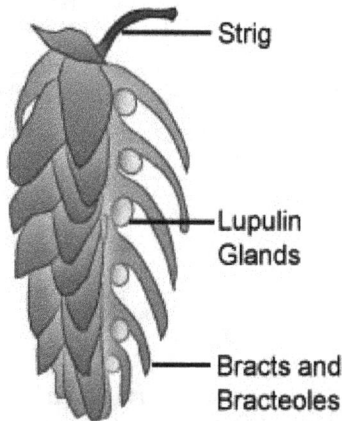

Figure 6.4.   Hop cone cutaway view.

Table 6.2. Raw hop composition.

Constituent	% w/w
Total resins	12–33
Essential Oils	0.2–3.0
Polyphenols	2–5
Proteins	15
Cellulose	40–50
Water	6–10
Monosaccharides	2
Pectin	2
Minerals	10

essential oils and the polyphenols define the brewing value of the hop cone.

The total resin is composed of a complex group of substances as depicted in Figure 6.5. According to the ASBC and EBC [6][7] the total resins are defined as that portion of the ether extract which is soluble in cold methanol. The total resin content ranges between 12 and 33 % and depends on the variety and the crop year. The total resins can be further divided into soft resins and hard resins depending on their solubility in hexane. The total soft resins are soluble in hexane, whereas the total hard resins are not soluble in hexane but in methanol. Some authors [6, 8] have investigated the different The most important components of the soft resin are the α- and β-acids, also known as humulones and lupulones respectively. The α- and β-acids are composed of different homologs with similar chemical structures. The main homologs composing the α-acid fraction are humulones, cohumulones and adhumulones. These three substances can represent up to 98–99% of the total α-acids (see Figure 6.6). Analogously, the main homologs representing the β-acids are known as lupulones, colupulones and adlupulones. Further structural variations exist and are referred to as post-, pre- and ad/pre-homologs but they are considered unimportant due to their low concentrations [1].

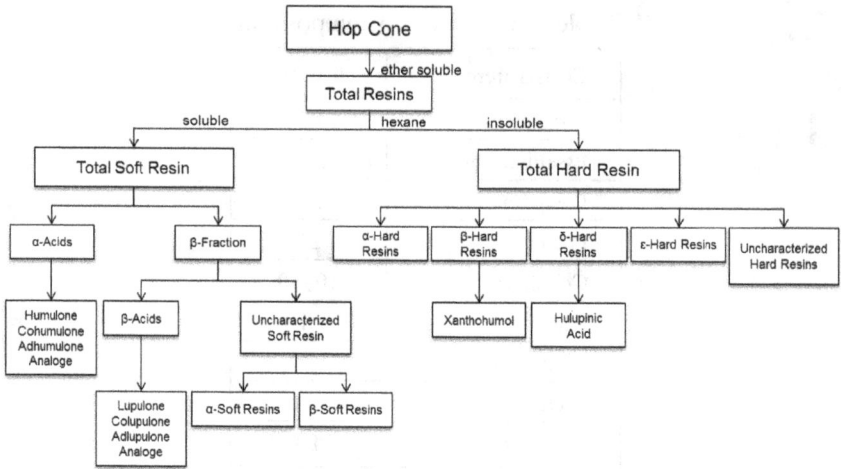

Figure 6.5. Total resin composition [6, 8].

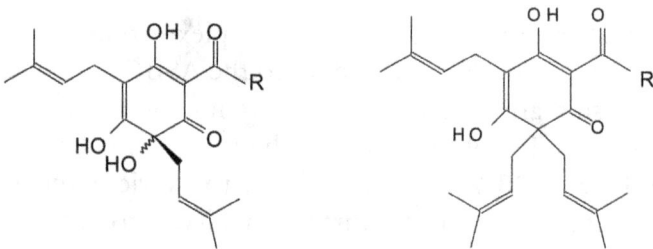

Humulone	-CH$_2$CH(CH$_3$)$_2$	Lupulone
Cohumulone	-CH(CH$_3$)$_2$	Colupulone
Adhumulone	-CH(CH$_3$)CH$_2$ CH$_3$	Adlupulone
Prehumulone	-CH$_2$CH$_2$CH(CH$_3$)$_2$	Prelupulone
Posthumulone	-CH$_2$ CH$_3$	Postlupulone
Adprehumulone	-CH$_2$-CH$_2$-CH$_2$-CH$_2$-CH$_3$	Adprelupulone

Figure 6.6. Chemical structure of the α- and β-acids.

The α- and β-acids are weak acids which means that they show a low solubility in acidic environments such as boiling wort (α-acids 3 ppm and β-acids almost zero). However, both acids are also important for preservation.

Table 6.3. Approximate proportions of various groups of compounds found in the essential hop oils [1, 9, 10].

Compound	Concentration
Monoterpenes	40 %
Sesquiterpenes	40 %
carboxylic acid esters	15 %
carboxylic acid	1 %
Monoterpene oxides	1 %
Sesquiterpene oxides	1 %
Aldehyde and Ketones	1 %
Aliphatic hydrocarbons	<1 %
Sulfur containing compounds	<0.1 %

The essential oils are volatile components that give hops their characteristic aroma. The total oil content of a hop variety may vary between 0.2 and 4.0% depending on the variety, the crop year and the growing region. By now more than 400 hop aroma compounds have been identified and characterized but it was suggested that there may be over 1000 different compounds [8][9].

Table 6.3 shows the approximate proportion of the most important groups of compounds found in the essential oil fraction. The monoterpenes, sesquiterpenes (both hydrocarbons) and carboxylic acid esters (oxygenated derivates) account for up to 95% of the essential oils.

The most important monoterpene is myrcene accounting for 30–60% of the oil content [10][11]. The aroma of myrcene is described as resinous, herbal and green. The major components that are present in the sesquiterpene fraction are β-caryophyllene and α-humulene. These two components account for up to 50% of the essential oils. Although myrcene, β-caryophyllene and α-humulene are present, the essential oils in higher concentrations compared to any other compound, due to the low solubility in water and wort as well as the evaporation during brewing, these compounds play only an important role if added during dry hopping.

Although the monoterpene oxides represent just 1% of the total oil content, one of its compounds, linalool, is considered as the characteristic impact compound that is responsible for a typical hop aroma. Linalool in beer is described to impart a floral aroma; however, its aroma contribution will depend on the hop addition. Its aroma threshold in beer is between 2–80 µg/L, normally defined as 10 µg/L.

Other monoterpene oxides are geraniol (rose-like aroma), nerol (floral aroma), citronellol (citrus aroma), α-terpineol (floral aroma), all of them having a relatively low threshold. Compounds that have thresholds of a factor 1000 lower than these (0.8–60 ng/L) are different thiols exhibiting different fruity flavors as 4-MMP (4-mercapto-methyl-pentan-2-one), 3 MH (3-Mercaptohexan-1-ol) and 3 MHA (3-Mercaptohexylacetate). The flavors are described as citrus, grapefruit, boxtree, passion fruit, black currant and also cat urine. So far it is known that the United States and Australian varieties contain higher amounts of free thiols as European varieties, which can partly explain the higher aroma intensity of these hop varieties [11][12].

Polyphenols contained in hops can be classified into phenolic carboxylic acids (e.g. ferulic acid), flavonols (e.g. quercetin), flavanols (e.g. catechin) and other polyphenolic compounds [12][13]. Although most of these substances are common in other plants, some polyphenols are exclusive to hops for example xantohumol.

In the hop plant, polyphenols are located mainly in the strig and bracts with the exception of the prenylflavonoids which are within the lupulin glands. Polyphenols account for 2–5% of the hop cone composition depending on the variety. They can be divided into low molecular weight and high molecular weight polyphenols. Low molecular weight polyphenols show an antioxidative capacity and increase the reduction power of the beer, which means that they protect the beer from oxidation increasing the taste stability. Moreover, low molecular weight polyphenols contribute to the mouthfeel of the beer.

High molecular weight polyphenols can cause an astringent bitterness, contribute to beer color and to haze formation, thus reducing the colloidal stability and causing turbidity of the beer.

6.6 Hop Addition in the Brewing Process

The presence of hop resins, essential oils and polyphenols in hops offers the possibility to the brewer to add bitterness and hop aroma to the beer, improve the microbiological, haze and flavor stability as well as the foam of the final product and contribute to the mouthfeel. Hops are added during the brewing process in either or both the kettle or after fermentation (see Figure 6.7).

The hop addition has to be adjusted to the desired effect in the beer. The primary role of hops in the brewing process is to impart bitterness to the beer. α-acids are almost insoluble in wort or beer and do not have an intense bitterness. If hops are added during boiling a chemical conversion, known as isomerization, occurs (see Figure 6.7). During isomerization the α-acids convert into a five ring structure, the iso-α-acids, which are more soluble and bitter than α-acids. Figure 6.8 shows the isomerization reaction of α-acids into iso-α-acids. The isomerization rate is defined as the percentage of the quantity of iso-α-acids measured in the wort divided by the quantity of α-acids added to the wort.

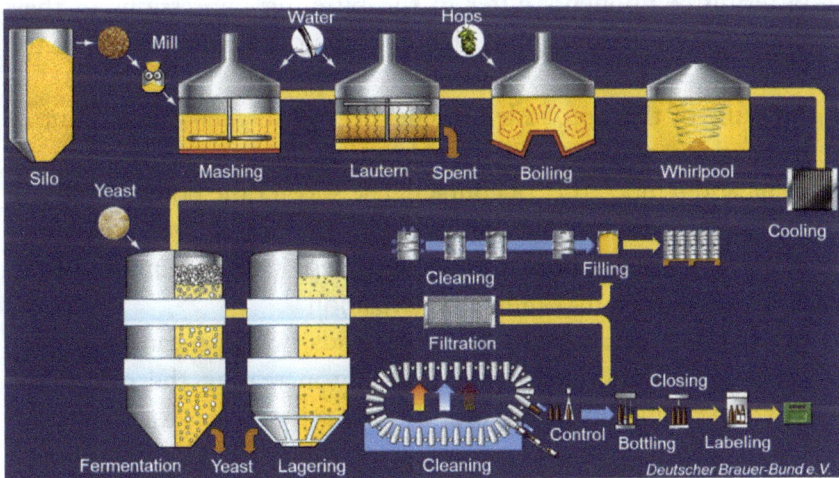

Figure 6.7. Schematic representation of the brewing process.

Source: Deutsche Brauer Bund e.V.

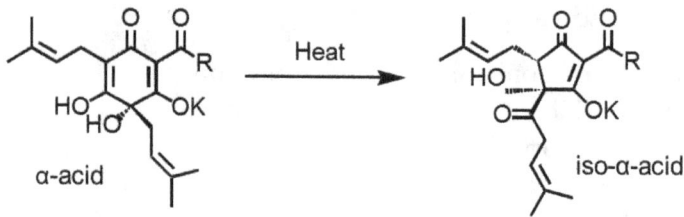

Figure 6.8.   Isomerization reaction of α-acids.

Isomerization of α-acids is influenced by the following parameters during the boiling process:

- Hop variety: There is a fluctuation in the composition of the different hop varieties. Depending on the variety chosen, the α-acid content is higher or lower. Also the crop year has an influence on the α-acid content.
- Type of hop product: There are products commercially available which offer a better utilization and/or higher (iso)-α-acid concentration (see Section, Hop Products).
- Freshness of the hop product: The degradation products of the bittering components differ in the bittering impressions of their not oxidized versions, thus changing the overall impression of the beer.
- Amount of the hop added: Generally, the higher the amount of hops added, the lower the isomerization yield.
- Time and temperature of contact: The shorter the boiling time and the lower the boiling temperature, the lower the utilization. Therefore, ideally, hops should be added at the beginning of the boil to achieve a maximum isomerization.
- pH of the wort: α-acids are weak acids, which means that they show low solubility at lower pHs. By increasing the pH of the wort, the solubility of the α-and iso-α-acids increases.
- Original gravity and composition of the wort: Generally, lower isomerization rates can be expected if brewing to higher gravity. Also the presence of substances in the wort that promote acidity or have a buffering capacity influence the isomerization rate.

- Type of wort boiling system: Kaltner, et al. [14, 15] studied the quality of the final beer using different boiling systems and found that depending on the boiling system used, the isomerization rate changes.

To account for the final concentration of iso-α-acids in a beer, not only the isomerization rate of the α-acids is of importance but also the utilization of the iso-α-acids during the brewing process. The utilization rate can be defined as the total amount of iso-α-acids in the final product related to the total amount of α-acids added. Utilization comprises all losses of α-acids and iso-α-acids during the brewing process until the finished beer. These losses include the isomerization reaction, α-acid degradation during boiling trub formation and precipitation, hot as well as cold trub, losses during fermentation (foam formation, yeast concentration), filtration and stabilization of the beer.

The following formula shows how to calculate the utilization rate of a product.

$$Utilization\,(\%) = \frac{iso - \alpha - acid\,in\,bear}{dosed\,\alpha - acid} \times 100$$

The final bitter intensity of a beer can be measured by analytical methods that determine the bittering units (BU or IBU) in the product. The Analytica EBC [14][16] as well as the ASBC Methods of analysis [15][17] describe the measurement of bitter units in beer. The hop resins are extracted with isooctane and the absorbance measured at 275 nm. The final result of the bitter units is calculated as follows:

$$IBU = abs275nm \times 50$$

However, this method is nonspecific and encompasses not only α-acids and iso-α-acids but also other bitter substances designated as auxiliary bitter (or not bitter) substances [1]. This method can be used as routine quality parameter in breweries but should not be used to compare beers. Also, it shows a poor correlation for downstream products or mixtures of hop products and it should not be used for the bitterness control of dry hopped beers. A specific method is available by means of HPLC. This method is able to quantify the amount of single bittering components.

Hops also have an influence on the final aroma of the beer. The hoppy aroma of a beer is achieved due to the addition of hop oils (in form of pellets, hop extract or specific hop oil products). However, most of the hop oils are poorly soluble into wort and are also volatilized during the boiling process. Of the 400 aroma compounds initially present, only a small portion remains in the wort, and their composition has been substantially altered [1].

To retain hop oils into the wort, and thus have a hop aroma in the final beer, hops should be best added at the end of the boil, during whirlpool or in the cold part of the brewing process (dry hopping).

Hop aroma in beer is a very complex issue in brewing. The following factors, among others, influence the hop aroma in the final beer:

- In the field:
  - Hop variety
  - Time of harvest and kilning
  - Crop year
- In the brewhouse:
  - Point of addition
  - The amount of hops
  - Solubility and evaporation properties of all compounds
- During fermentation:
  - Adsoprtion
  - Transformation interactions
- During lagering: If the beer is dry hopped
  - Dry hopping with or without the presence of yeast
  - contact time
  - Temperature
  - Agitation
- Filtration/pasteurization

The addition of hops for bittering is done according to the α-acid content of the product (g of α-acids/hL of beer). Therefore the total added quantity of the product can change between crop years depending on the α-acids content. However, addition of hop for aroma is more complicated due to the many factors influencing the final quality. Most brewers dose by grams of hops/hl, accepting that this goes along with varying amounts of vegetative, bitter and aroma compounds. Also the addition by the equivalent hop oil amount, e.g. 3 ml oil/hl (depending on variety this means 100–400 g/hl) is well established but still goes along with varying amounts of vegetative matter and bitter compounds. Some authors even recommend [1] to dose by specific compounds such as linalool, which is relatively easy to analyze in hops and beer. However, this goes along with varying amounts of total hop oil, vegetative matter and bitter acids. Newer research even questions the correlation of amount of hop oil and hop aroma intensity. Therefore, to obtain a consistent intense hop aroma in beer is quite a masterpiece and demands compromise. After all, hops is a natural product with variations from crop year to crop year. Different hop growing regions and even hop gardens show differences in the aroma of hops. Flavor science has shown that threshold values are only valid for single components. Synergistic, masking or additive interactions between hop oil components and other beer compounds lead to variations of the aroma in the final beer.

Due to the complex reactions that the hop oils undergo during the brewing process (some of them not completely understood) and that currently there is no valid approach for the determination of relevant flavor active compounds, sensory evaluation is the most accurate and most meaningful instrument to be used.

The Barth-Haas Group has developed a uniform tasting scheme with the help of flavorists and sommeliers for the assessment of hops and hoppy beers. This scheme works with 12 aroma categories identifying specific aroma attributes resulting in a defined and comparable aroma profile for the relevant hop variety or beer.

Figure 6.9 shows the Barth-Haas Group uniform tasting scheme. Each category should be rated in intensity from 1 to 10 (1 not perceived, 10 the highest). The specific attributes of each category to

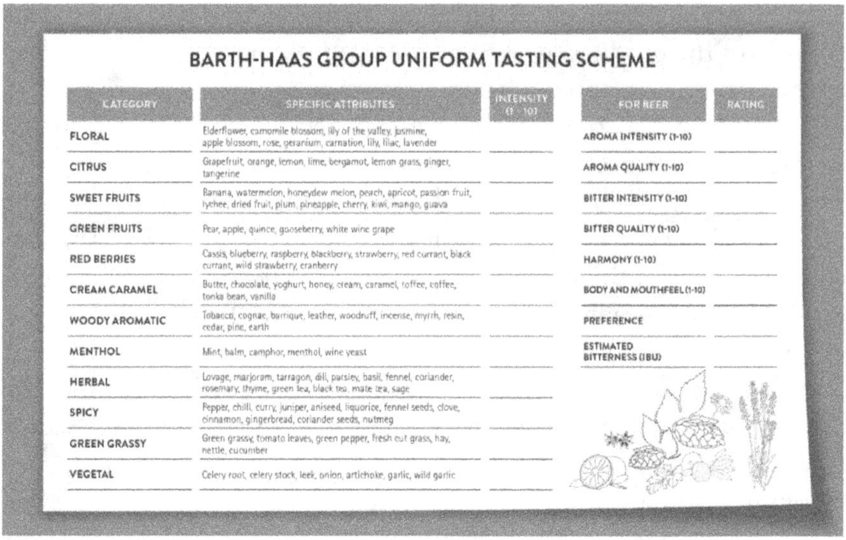

BARTH-HAAS GROUP UNIFORM TASTING SCHEME

CATEGORY	SPECIFIC ATTRIBUTES	INTENSITY (1–10)
FLORAL	Elderflower, camomile blossom, lily of the valley, jasmine, apple blossom, rose, geranium, carnation, lily, lilac, lavender	
CITRUS	Grapefruit, orange, lemon, lime, bergamot, lemon grass, ginger, tangerine	
SWEET FRUITS	Banana, watermelon, honeydew melon, peach, apricot, passion fruit, lychee, dried fruit, plum, pineapple, cherry, kiwi, mango, guava	
GREEN FRUITS	Pear, apple, quince, gooseberry, white wine grape	
RED BERRIES	Cassis, blueberry, raspberry, blackberry, strawberry, red currant, black currant, wild strawberry, cranberry	
CREAM CARAMEL	Butter, chocolate, yoghurt, honey, cream, caramel, toffee, coffee, tonka bean, vanilla	
WOODY AROMATIC	Tobacco, cognac, barrique, leather, woodruff, incense, myrrh, resin, cedar, pine, earth	
MENTHOL	Mint, balm, camphor, menthol, wine yeast	
HERBAL	Lovage, marjoram, tarragon, dill, parsley, basil, fennel, coriander, rosemary, thyme, green tea, black tea, mate tea, sage	
SPICY	Pepper, chilli, curry, juniper, aniseed, liquorice, fennel seeds, clove, cinnamon, gingerbread, coriander seeds, nutmeg	
GREEN GRASSY	Green grassy, tomato leaves, green pepper, fresh cut grass, hay, nettle, cucumber	
VEGETAL	Celery root, celery stock, leek, onion, artichoke, garlic, wild garlic	

FOR BEER	RATING
AROMA INTENSITY (1-10)	
AROMA QUALITY (1-10)	
BITTER INTENSITY (1-10)	
BITTER QUALITY (1-10)	
HARMONY (1-10)	
BODY AND MOUTHFEEL (1-10)	
PREFERENCE	
ESTIMATED BITTERNESS (IBU)	

Figure 6.9. Barth-Haas Group uniform tasting scheme.

help the assessor to identify the exact aroma. Also the aroma intensity and quality, the bitter intensity and quality as well as the harmony, the body and mouthfeel can be rated from 1 to 10. Lastly, to compare between beers, the preference can be ranked and the bitter units can be estimated.

6.7 Hop Products

Hops can be used in many forms for beer production. The simplest form is known by a number of different names which include whole hops, raw hops, cone hops, leaf hops or baled hops [16][18]. However, a vast array of other hop products is available, all produced from whole hops by specialized technologies and used for numerous applications in various ways.

The main objectives of hop processing are:

- Improved handling:
  - Reduction of the volume of the product: for example, the density of raw hops is 100–150 kg/m^3 compared to that of pellets 480–550 kg/m^3 or that of CO_2 extract 1,000 kg/m^3

- Flowability: Some products become flowable when they are produced
- Improve stability: Hop products are protected from oxidation in special packaging and therefore, some products are stable for longer time
- Improve homogeneity (e.g. α-acid content): Due to blending during processing which also facilitates addition of the exact quantity of product
- Increase utilization: Depending on the product and how it is used, the α utilization can be up to 90%

Figure 6.10 shows the hop product classification. Hop cones, as the simplest form of products are dried on the farm and arrive to the hop marketers baled where they are normally packed in gastight foil bags and stored cold. Although hop cones comprise a number of disadvantages compared to other products (bulkiness, not homogeneous, provide poorer utilization among others), there are some breweries that are still using them. However, just about 2–3% of the hop products used are hop cones.

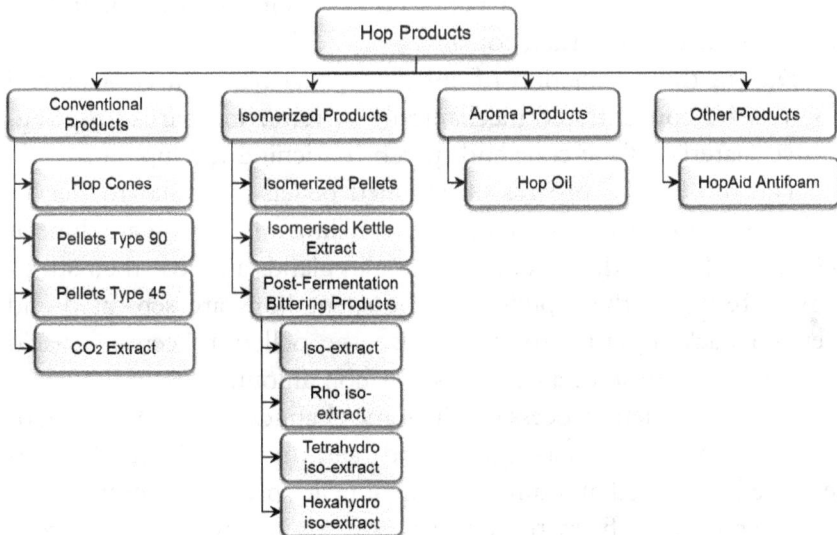

Figure 6.10. Classification of hop products.

The vast majority of products used in the brewing industry are pellets. Generally, there are three types of pellets: normal pellets (type 90), enriched pellets (type 45) and isomerized pellets. The production process steps of normal pellets are as follows (see Figure 6.11):

- Bale opener
- Homogenization of the batches
- Rotary magnet and gravity separator for elimination of foreign materials and other parts of the plant
- Drying to a moisture content of 8–10%
- Milling
- Powder mixer
- Pelletization
- Sieving
- Cooling
- Weighing, filling and packaging in foil bags

Although normal pellets have the same composition as hop cones, they show some advantages: (i) the stability of the product is increased due to the nature of the pellet and the protection against oxygen in the packaging, (ii) they show better homogeneity and (iii) the utilization is slightly higher than cones.

During the production of enriched pellets, the bitter acids and hop oils are concentrated mechanically by removing part of the vegetative material of the cone and thereby enriching the lupulin.

The production process of enriched pellets is similar to that of pellets type 90 with the exception that milling and sieving are done at −35°C (−31°F) to deep freeze the lupulin glands and lose their stickiness. Afterwards the lupulin and the spent hops are separated and blended back together to standardize the pellets to certain α-acid content by addition of a defined spent hop amount.

The enrichment process entails some changes in the total composition. In enriched pellets, the α- acids and the hop oils can be concentrated to a fixed amount in the final pellet product chosen by the customer ranging from the original content of the cone hops to a maximum content (e.g. 4% α up to a maximum of 20%). Oppositely, due to the removal of some vegetative material, the content of

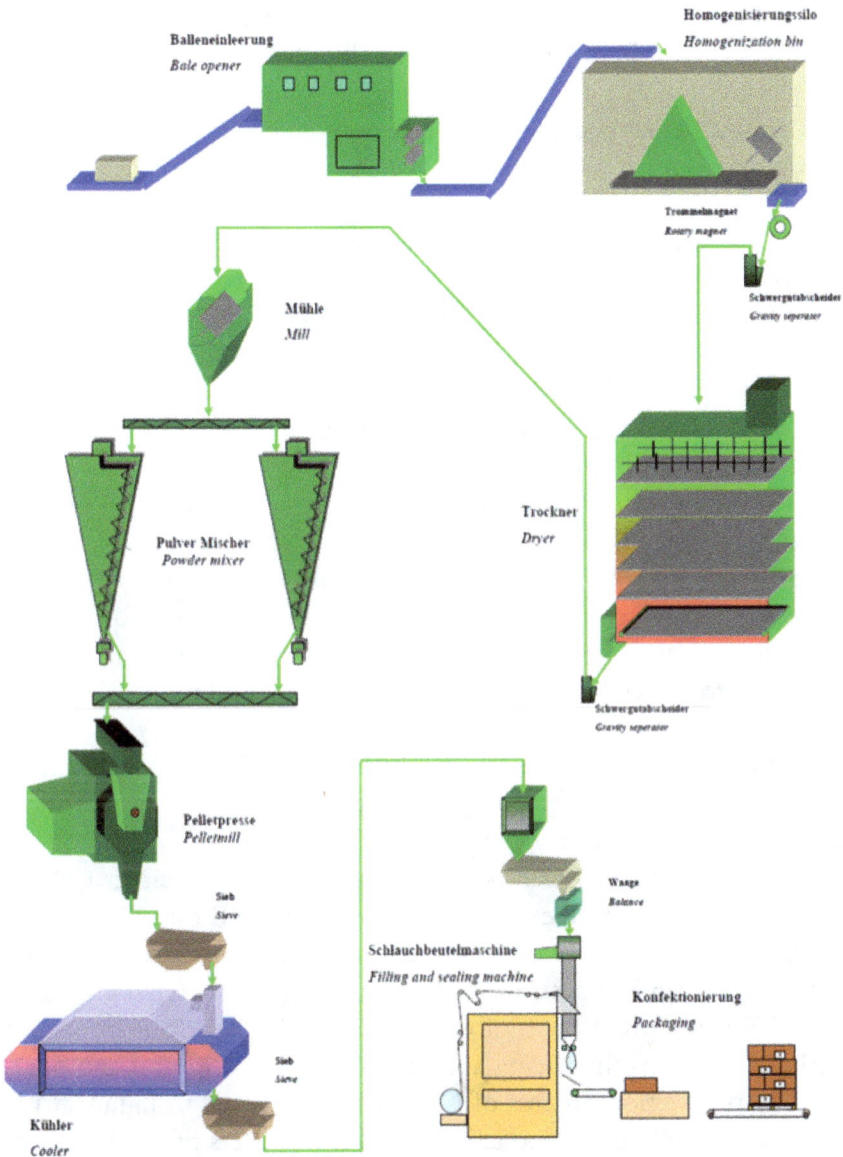

Balleneinleerung
Bale opener

Homogenisierungssilo
Homogenization bin

Trommelmagnet
Rotary magnet

Schwergutabscheider
Gravity seperator

Mühle
Mill

Pulver Mischer
Powder mixer

Trockner
Dryer

Schwergutabscheider
Gravity seperator

Pelletpresse
Pelletmill

Sieb
Sieve

Waage
Balance

Schlauchbeutelmaschine
Filling and sealing machine

Konfektionierung
Packaging

Sieb
Sieve

Kühler
Cooler

Figure 6.11.   Production process steps of normal pellets.

polyphenols is reduced accordingly, with the exception of xantohumol as this polyphenol is present in the lupulin glands.

Enriched pellets have some advantages compared to cones or normal pellets. They provide improved homogeneity, the storage and transport costs are reduced, and they show lower wort losses due to a smaller percentage of hot break material.

The production process of isomerized pellets is similar to that of pellets with the difference that 1–3% of MgO (depending on the original α-acid content) is added to the powder mixer and once the pellets are packed they are kept in hot rooms at 40–50°C for 10–14 days. After this period of time, the foils are cooled down. The addition of MgO and the following hot temperatures, force the α-acids to isomerize to iso-α-acids within the pellets.

The composition of isomerized pellets is the same as pellets, depending how they have been produced, with the exception that α-acids are converted to iso-α-acids. The prior isomerization present some advantages compared to the other pellets or hop cones. The iso-α-acids present in isomerized pellets are more soluble into wort and therefore need less contact time during boiling and achieve higher utilization rates.

All the above mentioned pellets have α-acids and hop oils so they can be used for bittering (boiling addition) as well as for aroma addition (whirlpool or dry hopping addition).

Normal hop pellets can be extracted with solvents, either CO_2 or ethanol, as both are able to dissolve hop resins and oils. CO_2 under super critical conditions (temperature and pressure above the standard temperature and pressure) adopts properties between a gas and a liquid. For the production of CO_2 extract, CO_2 is held under super critical conditions thus being able to extract the hop resins and oils.

The production process of CO_2 extract is shown in Figure 6.12. It consists of filling the extraction vessel with pellets type 90 and letting CO_2 under super critical conditions to pass through. Afterward there is a release of the pressure, the brewing component fall in the separator and the CO_2 is recovered. The process is not continuous, the extraction vessel and the separator have to be emptied and filled again to run another batch.

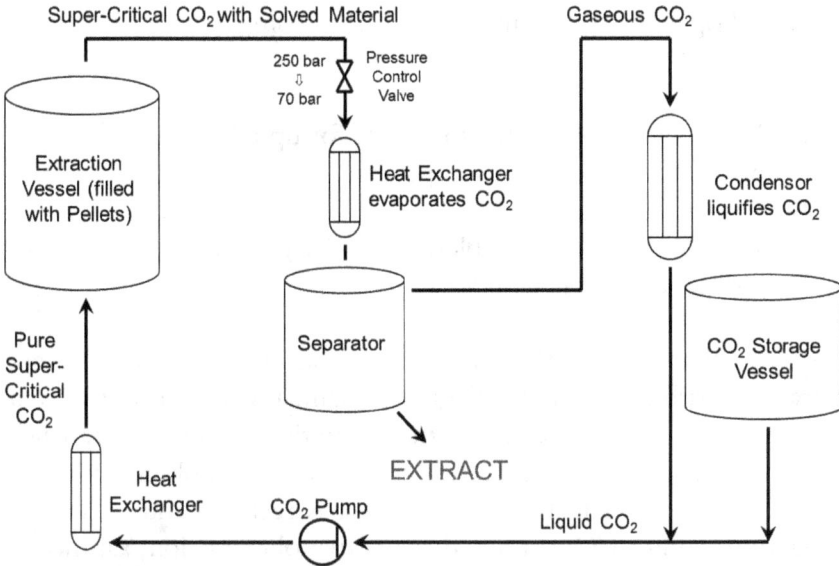

Figure 6.12. Production process of CO_2 extract.

Table **6.4.** Average composition of pellets type 90 and CO_2 extract.

% w/w	Pellets	CO_2 Extract
α-Acids	2–20	35–60
Oil Content	0.5–3.0	3–12
α-Acids	1–10	15–40
Polyphenols	2–5	0
Water	6–10	0

CO_2 hop extract contains the α-and β-acids, essential oils and other soft resin components of hops concentrated by 4–5 fold. The α-acid contestation ranges from 30–65% depending on the variety extracted. However, during the production process the polyphenols are not extracted as they are not dissolved by CO_2. Table 6.4 shows the average composition of CO_2 extract compared to that of pellets type 90.

CO_2 hop extracts have many advantages compared to other kettle products:

- Higher stability: CO_2 extract is stable for up to 8 years
- Slightly better utilization than pellets
- Better homogeneity
- Reduced levels of undesirables, particularly inorganic compounds i.e. nitrates and some pesticides

The ethanol extraction involves a continuous countercurrent extraction with fermentation alcohol (ethanol) serving as the solvent [1]. The extract is obtained through evaporation of the solvent, which is then separated by centrifugation into a water-soluble fraction and the pure resin extract. The alcohol vapor is rectified and reused for extraction. Unlike CO_2 extract, during ethanol extraction, xanthohumol and other prenylflavonoids are almost completely extracted, nitrate is significantly reduced and the pesticide residues are either dissolved or reduced.

Isomerized kettle extract, also called IKE, is produced by heating CO_2 extract with $MgO/MgCO_3$. The presence of a catalyst isomerizes the α-acids to iso-α-acids with a greater than 95% transformation. It contains almost the same ingredients as CO_2 extract (from the same variety) with the exception that the α-acids are isomerized. IKE offers an alternative to CO_2 extract resulting in higher bitterness efficiency by replacement of hops, pellets or extract in the kettle. IKE may be added at any time during wort boiling, as the iso-α-acids dissolve rapidly. Due to the prior isomerization of the product, there is a significant increase in the utilization (50–60%).

The next group of products (iso-extract, tetrahydro iso-extract, rho iso-extract and hexahydro iso-extract) is classified as post-fermentation bittering products. Iso-extract is an aqueous alkaline solution of the potassium salts of iso-α-acids standardized to 30% w/w by HPLC. Iso-extract is produced by isomerization of α-acids using alkali and/or divalent cations. It improves utilization by replacing kettle hops or it can be used to adjust bitterness in beers that have been under hopped in the kettle.

Tetrahydro iso-extract, rho iso-extract, and hexahydro iso-extract are also known as hydrogenated or reduced iso-extracts due to the disruption of the double bonds with hydrogen. This hydrogenation prevents the formation of 3-methyl-2-butene-1-thiol, the key aroma compound responsible for sunstruck flavor in beer.

Post-fermentation products have shown some advantages compared to other extracts or products used for bittering:

- Better utilization: up to 85% of utilization if used post-fermentation.
- Foam and lacing enhancement: Tetrahydro iso-extract and hexahydro iso-extract have shown a better foam and lacing in beer.
- Improved light stability: Tetrahydro iso-extract, rho iso-extract, and hexahydro iso-extract are light stable but iso-extract is not.
- Different bittering impression: The bittering impression of iso-extract is defined as clean and non-lingering, of rho iso-extract as clean and smooth, of tetrahydro iso-extract as harsh and lingering and of hexahydro iso-extract as harsh but less than tetrahydro iso-extract.

Hop oils are considered post-fermentation aroma products as they are able to impart just aroma to the beer. They are extracted from CO_2 extract by means of steam distillation or molecular distillation. Hop oils can impart a hoppy aroma but to maintain the aroma in the final beer they need to be added late in the process. However, hop oils are not soluble in cold wort or beer and need to be mixed with an emulsifier or a carrier such as propylene glycol or ethanol.

The Barth-Haas Group offers also a product called HopAid Antifoam® which is aim to be used during fermentation to prevent excessive foam formation. This product is the emulsion of a hop extract in water and offers the possibility of increase in the fermenter utilization which returns in lower unit costs of production.

6.8 Topics for Discussion

- Discuss the importance of environmental factors in determining which strain of hops would be suitable for cultivation in a specific location

- How does the level of β-acids impact selection of hop variety and hop addition rate?
- Discuss the effects of timing on the addition of different types of hops to the boil and their effects on beer characteristics
- How does the practice of "dry hopping" affect the impact of bittering hops added to the boil?
- What types of novel products from hops can you think of that may have a commercial potential?

6.9 List of References

1. Biendl M, Engelhard B, Forster A, *et al. Their Cultivation, Composition, and Usage.* Hans Carl Verlag; 2014.
2. Mahaffee WF, Pethybridge SJ, Gent DH. *Compendium of Hop Diseases and Pests.* Amer Phytopathological Society; 2009.
3. Eßlinger HM. *Handbook of Brewing.* Darmstadt: WILEY-VCH Verlag GmbH & Co. KGaA; 2009.
4. Meier H. *The Barth-Report 2017–2018.* Nürnberg: Joh. Barth & Sohn GmbH & Co KG; 2018.
5. Roberts TR, Wilson RJH. Hops. In: *Handbook of Brewing.* Boca Raton, FL: Taylor & Francias; 2006, pp. 177–280.
6. Stevens R. The chemistry of hop constituents. *Chem Rev.* 1967;67: 19–71.
7. ASBC and EBC. Hop resin nomenclature. *J Inst Brewing.* 1957;63: 268–288.
8. Palamand SR, Aldenhoff JM. Bitter tasting compounds of beer. Chemistry and taste properties of some hop resin compounds. *J Agric Food Chem.* 1973;21:535–543.
9. Roberts MT, Dufour JP, Lewis AC. Application of comprehensive multidimensional gas chromatography combined with time-of-flight mass spectrometry (GCxGC-TOFMS) for high resolution analysis of hop essential oil. *J Sep Sci.* 2004;27:473–478.
10. Hops Academy Seminar. Joh. Barth und Sohn GmbH & Co KG; 2019.
11. Almaguer C, Schönberger C, Gastl M, Arendt EK, Becker T. Humulus lupulus — a story that begs to be told. A review. *J Inst Brewing.* 2014;120:289–314.

12. Roland A. A powerful analytical indicator to drivevarietal thiols release in beers: the "ThiolPotency." *Brewing Sci.* Nov/Dez 2017.
13. Biendl M, Pinzl C. *Hops and Health. Uses — Effects — History.* Wolnzach: German Hop Museum Wolnzach; 2008.
14. Kaltner D. Moderne Aspekte zur Hopfung des Bieres. *Brauindustrie.* 2005; 90(2):22–25.
15. Kaltner D, Mitter W, Binkert J, Preis F, Zimmermann R, Binedl M. SchoKo wort boiling system and hop components. *Brauwelt Int.* 2006;I:40–43.
16. Analysis Committee of the EBC. Analytica-EBC: European Brewery Convention.
17. American Society of Brewing Chemists, ASBC Methods of Analysis.
18. John I. Haas, Inc. *Barth-Haas Hops Companion.* Yakima, Washington: John I. Haas, Inc.; 2013.

Chapter 7

Water Quality and Usage

7.1 Importance

Most breweries have either three or four classifications of water based on their use: process water (as used in the mash to make product), boiler feed water, cleaning water and cooling water. Wastewater is from a mixture of sources but mostly consists of contaminated cleaning water and disposal of product and intermediates such as trub and spent yeast. In small breweries, process water and cleaning or utility water are usually of the same quality. There are both economic and environmental reasons for minimizing water usage in the brewery. Not only are volumes high, on average four to five times and sometimes more than seven times as great as the volume of beer produced, but depending on the size of the brewery, up to 70% of this water may require some form of treatment prior to discharge to the municipal system.

The full cost of water includes not only the initial purchase from the municipal supply but also any costs for upgrading the water quality (for example, filtration and softening), wastewater pretreatment and/or a sewer service charge and any labor associated with maintaining these operations. Water conservation not only makes good economic sense but may be critical to ensure year-round operation in certain locales. Water shortages are becoming commonplace, both in the United States of America and around the world, and are likely to get worse as global warming continues. Strategies for minimizing water usage in the brewery are addressed in Section 7.8.

7.2 Composition

Water is the largest single component of beer and, therefore as would be expected, the composition of the water is a key factor in determining the characteristics of the product. High purity water in which the presence of trace elements has been almost entirely eliminated, is not desirable for brewing applications. Indeed, the mineral content of the water can have numerous beneficial effects on the brewing and fermentation processes. However, the composition of water as supplied to the brewery is highly dependent on the brewery location, as both the surrounding geology and design of the municipal water distribution system will have a major impact. So, although it is difficult to define what "typical" brewery water may consist of, in general it will contain a mixture of inorganic ions, both cations (positively charged) and anions (negatively charged) in solution at a total concentration between 0.5 and 1.5 g/L. In addition to these ions, there will be varying amounts of undissolved solids, contributing a level of turbidity, and trace amounts of both natural organic material and unwanted pollutants such as fertilizers and pesticides. The concentrations of these various components can affect enzymatic activity during mashing, the yeast activity during fermentation and also characteristics of the final product, including flavor and foam stability.

Historically, breweries were often established based on the characteristics of the local water supply, thus beer styles became associated with different geographical regions (see Table 7.1), the bitter ales from the United Kingdom and the lighter lagers from central and eastern Europe. Fortunately, this limitation is no longer a barrier and

Table 7.1. The inorganic ion concentrations of water associated with select beer styles.

Location (beer style)	Na^+	Mg^{2+}	Ca^{2+}	Cl^-	SO_4^{2-}	HCO_3^-
Burton (UK pale ale)	54	24	352	16	820	320
Pilsen (Czech lager)	32	8	7	5	6	37
Dublin (Irish stout)	12	4	119	19	54	319

modern technology can be used to alter the composition of the water to suit the needs of the brewer.

7.3 Hardness and pH

The salt content of the water exists in the disassociated forms of positively and negatively charged ions, the most important for brewing being the bicarbonate ion, HCO_3^-. This source of alkalinity has a large impact on the changes in pH that occur during mashing and boiling. Water hardness can be both "temporary" or "permanent" depending on which salt is responsible. Calcium bicarbonate, $Ca(HCO_3)_2$, is considered as temporary hardness because on boiling it forms a hard precipitate, $CaCO_3$, and releases CO_2 gas. Permanent hardness is based on the presence of other salts such as $CaSO_4$ (gypsum), $CaCl_2$ and $MgSO_4$, which do not precipitate on boiling. The free hydrogen ion concentration of water is a function of the relative amounts of bicarbonate ions, HCO_3^-, and carbonic acid, H_2CO_3.

The bicarbonate ions exert alkalinity by removing free hydrogen ions from the water to form carbonic acid. Thus, the higher the bicarbonate concentration, the higher the level of alkalinity and associated pH. Heating of the mash will disassociate the carbonic acid into CO_2 and water, further increasing pH, or depending on the level of calcium, form the insoluble carbonate, $CaCO_3$. The buffering action of the bicarbonate ion–carbonic acid equilibrium around a pH of 6.4 is illustrated in Figure 7.2.

As a pH of 6.4 is too high for optimal enzyme activity during mashing, fortunately the bicarbonate ion is not the only factor determining wort pH. During mashing, phytase enzymes that occur naturally in the barley, release bound phosphates from the phytin salts, the main reserve of phosphate in the barley. The phosphates are present in the wort as two potassium salts, the monobasic form KH_2PO_4 and the dibasic form K_2HPO_4. The relative quantity of these two salts will largely determine the buffered pH in the mash. With little or no alkalinity present (low level of bicarbonate), the typical mash pH will be close to 5.4. However, the presence of calcium bicarbonate,

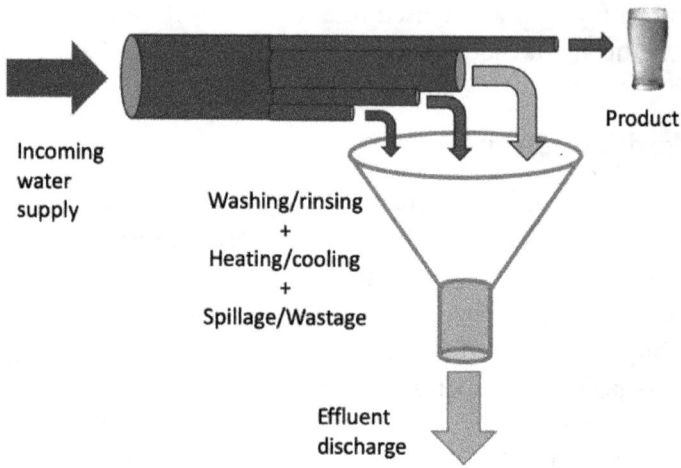

Figure 7.1. Water usage in the brewery—a relatively small proportion of the total water supply actually ends up in the product.

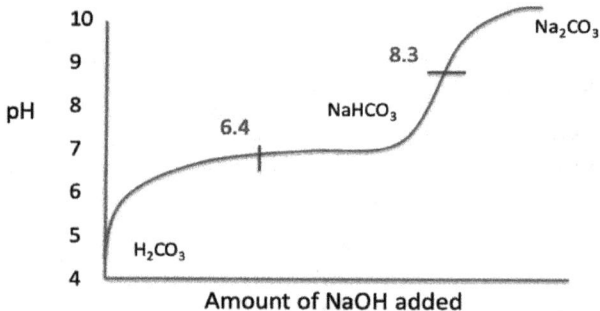

Figure 7.2. A titration curve for carbonic acid with sodium hydroxide, showing the strong buffering effect around the pH of 6.4.

$Ca(HCO_3)_2$, will exert alkalinity by the formation of insoluble calcium phosphate, $CaHPO_4$, from the acidic KH_2PO_4, resulting in an increase in pH, the extent depending on the level of alkalinity in the mash water.

$$Ca(HCO_3)_2 + 2KH_2PO_4 \rightarrow CaHPO_4 + K_2HPO_4 + 2H_2O + 2CO_2$$

Excess alkalinity resulting in a high wort pH has mostly detrimental effects on brewing. These include:

1. Decreased activity of both the amylase and protease enzymes, adversely affecting both sugar yield and level of free-amino nitrogen in the wort
2. Increased wort viscosity, adversely affecting lautering
3. Reduced phytase activity and phosphate availability
4. Increased formation of polyphenols and astringency
5. Reduced color formation from malt constituents
6. Increased haze formation as a result of higher levels of β-glucans, tannins and stabilization of colloidal protein

The one positive effect of elevated pH is the enhanced bittering effect from the hop acids and the associated reduction in hop usage

If necessary, the pH of the mash can be adjusted by chemical addition. Alkalinity can be directly neutralized with either sulfuric, hydrochloric, phosphoric or lactic acid. However, addition of these acids will result in the production of neutral salts, which can adversely affect beer taste. Depending on the quantity of acid needed and the style of beer being produced, this may not be a desirable approach. Alternatively, the addition of either $CaSO_4$ or $CaCl_2$ to the mash water will counteract the alkalinity by decreasing the concentration of the dibasic potassium salt, precipitating the calcium and lowering the pH.

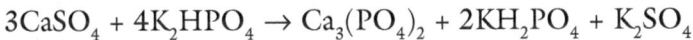

$$3CaSO_4 + 4K_2HPO_4 \rightarrow Ca_3(PO_4)_2 + 2KH_2PO_4 + K_2SO_4$$

The quality of lager beer is particularly sensitive to the presence of excess minerals, so a technique based on the addition of lime, $Ca(OH)_2$, can be used to avoid the production of neutral salts. The calcium carbonate will precipitate during the boil.

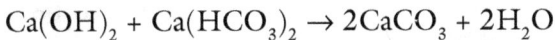

$$Ca(OH)_2 + Ca(HCO_3)_2 \rightarrow 2CaCO_3 + 2H_2O$$

However, with lime addition, care must be taken not to overdose as excess lime will actually increase the pH.

Table 7.2. Major cations in wort and their potential effects on fermentation and beer quality.

Cation	Ideal range in wort after boil (mg/L)	Potential effects
Calcium	40–100	Yeast growth, flocculation, foam stability
Magnesium	Less than 15	Can enhance bitterness and astringency
Sodium	75–150 max.	Can enhance mouthfeel
Potassium	Less than 10	Can be perceived as salty
Zinc	Less than 0.2	Yeast growth, can be metallic
Iron	Less than 0.2	Mouthfeel and perceived as metallic

Associated with the anionic bicarbonate ion, the most important cation is calcium. Sufficient calcium is necessary for both amylase and protease activity, reducing wort haze by combining with oxalates and proteins to form trub during the boil, and as an essential element for yeast growth and flocculation. Other less important cations include magnesium, sodium, potassium, zinc and iron as shown in Table 7.2.

Various anions can also have important effects in the wort, especially chloride and sulfate, both of which should be less than 250 mg/L. Silicates and nitrates are generally undesirable and should be limited to a maximum of 25–30 mg/L.

7.4 Water Pretreatment

Process water is high quality water that is used to perform the mashing process and therefore ultimately comprises about 95% of the final product. Process water is sourced either from a municipal supply or from a private well and the quality can be variable, even seasonally, based on source. A number of different treatment options are available as shown in Figure 7.3, however, it is not a general practice to include a de-ionization step, unless the water is to be reconstituted differently for each beer style.

Almost all water originating from either municipal systems or private wells will have some level of insoluble particulates. To avoid these

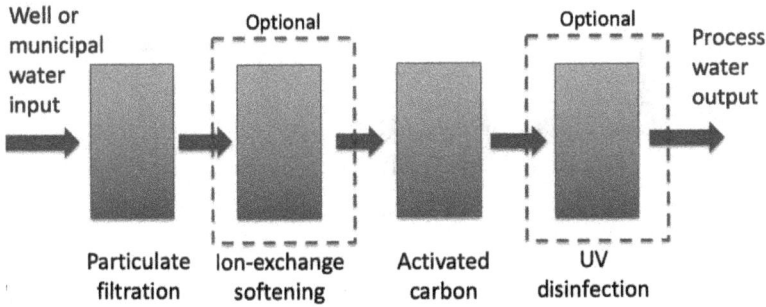

Figure 7.3. A sequence of steps that could be employed for the treatment of brewery process water. Incoming water quality will determine the requirements for softening or disinfection.

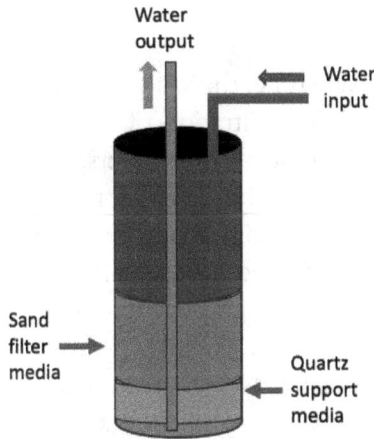

Figure 7.4. A high throughput pressurized sand filter for removing particulates from the water supply.

particulates from either fouling subsequent treatment steps or entering the brewing process, some type of filtration should be employed.

For relatively large volumes of water, sand filters are commonly used. As illustrated in Figure 7.4, the incoming water percolates under pressure through a layer of fine sand supported by coarse support media. The sand traps the suspended particulates and allows the clarified water to be piped out from underneath the support layer. As the quantity of

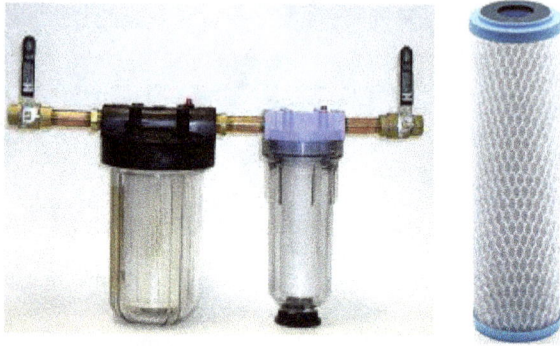

Figure 7.5. A series combination of cartridge filters, the larger filter is designed for particulate removal and the smaller cartridge is composed of activated carbon as pictured on the right.

particulates builds up over time, the sand needs to be either back flushed or periodically replaced. The filtration of smaller quantities of water is more efficient with the use of cartridge filters, often configured as a pair as pictured in Figure 7.5, the first cartridge made from cellulose for particulates and the second cartridge made from activated carbon for dissolved organics and chlorinated compounds. The effectiveness of the activated carbon is due to its high porosity and surface area for adsorption. The carbon may be either granular (GAC type) or in a powdered form (PAC type). It is possible to regenerate activated carbon by heating, but usually both cartridges are replaced on a periodic maintenance schedule and the old cartridges disposed of.

Softening of the water may be desirable in locales with high levels of minerals in the water supply. Softened water helps to reduce scale buildup in boilers and also increases the effectiveness of cleaning agents. The most common system employed for softening is based on ion exchange. Negatively charged plastic zeolite beads are packed in a column and exposed to a brine solution with a high concentration of positive sodium ions that are attracted to the beads. As the incoming water passes through the column, the more soluble sodium is displaced from the beads by calcium and magnesium ions, thereby removing temporary hardness from the water. The beads need to be periodically regenerated by washing with the brine solution, displacing the calcium and magnesium ions from the beads and sending them to the drain. Although ion-exchange softening has advantages for both boiler feed water and wash water, it may not be desirable for brewing water, as calcium has important functions and increased levels of sodium may adversely affect beer flavor.

Disinfection of process water, for example with ultraviolet light, is usually not necessary as municipal water systems perform their own disinfection, typically by chlorination or ozonation to eliminate any potential pathogens. If private well water is used, then it should be tested to ensure that it does not contain any coliform bacteria that could have contaminated the well, for example, from agricultural operations. The boiling step, as part of brewhouse operations, should normally be sufficient to sterilize the wort unless the process water has very high levels of bacteria. As described in the subsequent section, the use of antimicrobial sanitizers in the rinse water as part of a cleaning and sanitation strategy for vessels and pipes, should also eliminate the need for prior disinfection of the utility water.

7.5 Cleaning and Sanitation

For a brewer to maintain consistent high quality of product batch after batch, it is critical that a proper regime of cleaning and sanitation be followed. Thorough cleaning of vessels, pipes and hoses minimizes both residue carryover from previous batches and helps prevent the formation of microbial colonies that could ultimately contaminate the process, resulting in off-flavors and beer spoilage. By its nature, beer is resistant to bacterial contamination, largely due to its ethanol content, low pH, presence of hop acids and lack of residual sugar. However, brewer's wort has no ethanol, is only moderately acidic and has a large amount of sugar and other nutrients that are very attractive to bacteria and wild yeast. Following the boil, unless the pipes, hoses, heat exchanger, aerator and fermenter have all been properly cleaned and sanitized, contamination of wort can easily occur during transfer from the kettle.

Although many microbes can be found in the brewing environment, typical contaminants can include wild yeast (various

Saccharomyces sp. and *Brettanomyces sp.*), *Lactobacillus brevis* and *plan-tarum*, *Pediococcus damnosus* and *Acetobacter aceti*. The effects of microbial contamination include: reduced yield of ethanol, haziness, sourness (from lactic and acetic acids), increased attenuation (from the wild yeast *S. diastaticus*) and off-flavors.

Although manual cleaning of some smaller vessels and equipment can be effective, the use of clean-in-place systems (CIP) provides numerous advantages with larger vessels (see Figure 7.6). Once a CIP process has been properly validated for a specific vessel, unless there are significant changes to the usage of the vessel, the cleaning will be carried out thoroughly and efficiently time after time.

Detailed specifications for a CIP process are dependent on the characteristics of both the vessel and the material that is being processed in the vessel. However, most CIP processes will consist of the following steps:

1. Initial rinse to drain
2. Recirculated hot caustic wash

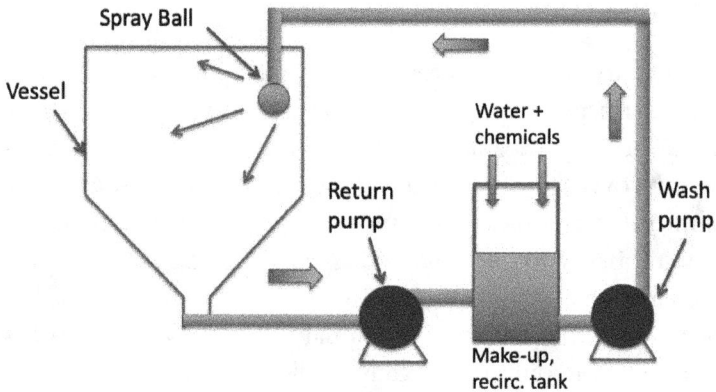

Figure 7.6.   A typical CIP setup consisting of two pumps and a make-up or recirculation tank. Many systems also incorporate a heat exchanger, sensors for temperature and conductivity and flow meters. Automated systems include data logging that provides a record of when each cleaning was performed. Use of separate recirculation tanks for each step, allows for reduced water consumption.

3. Intermediate rinse at ambient
4. Recirculated hot acid sanitizer
5. Final rinse to drain with process water

To achieve an acceptable level of residuals, during CIP development, the recirculated wash and sanitizer steps are carried out for different lengths of time at specific temperatures before the CIP process is finalized. Typical ranges for the wash steps would be between 5 and 15 minutes at 85–95 °C. Other CIP variables include: types of chemicals and concentration used in wash and sanitizer solutions, minimum flow rates and pressures needed to ensure spray ball performance and the design of the equipment being cleaned, especially surface finish. Automated CIP systems use a programmable logic controller (PLC) to ensure that each step is carried out to specification and to log pertinent data such as: the duration of each step, temperatures, pH or conductivity of both supply and return solutions, supply pressure and flow rates. This data is an important record that may be needed when investigating possible sources of process contamination.

The specific composition of organic residues encountered in brewery vessels and pipes are dependent on their particular use, whether it be for mashing, boiling, cooling, fermentation or maturation, but will typically include various amounts of carbohydrates, proteins, lipids, polyphenols, hop acids and insoluble grain, hop and yeast solids. In addition, inorganic residues such as calcium oxalate combine with proteins to form "beer stone." This is a white residue that can form on surfaces downstream from the kettle and if not removed by the caustic wash, over time will be an ideal environment for bacterial colonies to survive from the effects of the sanitizer.

Specialized caustic detergents have been designed for removal of brewing residues, usually based on a source of alkali such as sodium hydroxide, potassium hydroxide or sodium silicate with added non-foaming surfactants and water conditioners. However, as with any CIP process, the specific parameters that are necessary to achieve effective removal must be validated under actual brewing conditions. The sanitization step follows the caustic wash and is usually based on a mild acid combined with an oxidizer. The low pH of the sanitizer will help to remove both residual alkaline residues from the previous wash and mineral residues, while also providing a degree of microbial disinfection. An example of a commercial sanitizing solution includes peroxyacetic acid in combination with hydrogen peroxide. Sodium hypochlorite and iodine are also effective oxidizers with disinfection properties.

7.6 Wastewater Treatment

Brewery wastewater is generated from a variety of sources and is characterized by high variability depending on what activities are the focus of attention in the brewery. In general, wastewater is described based on certain specific characteristics: five-day biochemical oxygen demand (BOD_5), suspended solids (total–TSS and volatile–VSS), ammonia nitrogen and phosphate. Each operation within the brewery generates wastewater that differs in the concentrations of these parameters and, when combined with the relative volume of water generated, contributes a different proportion to the overall discharge load that must be treated by the municipality. Table 7.3 describes the principal brewery operations and associated wastewater generated (adapted from Water and Wastewater: Treatment/Volume Reduction Manual, Brewers Association). Municipal sewerage fees or taxes can vary significantly based on the location of the brewery, the size of the municipality in relation to the brewery discharge volume and the operation of the brewery, especially regarding water management strategies. The municipality may require continuous monitoring of the brewery effluent and base the sewerage fees on specific criteria, such as both average and peak flow, average BOD_5 and suspended solids levels and pH range.

In the United States of America, no federal regulations on pretreatment standards are currently applied specifically to the brewing industry. However, depending on the state and local municipality, some pretreatment requirements may still be enforced. Typically, the most stringent standards apply to those companies that are considered as "Significant Industrial Users" or SIU. As a guideline, an SIU may be considered as an operation that produces greater than 25,000

Table 7.3. The primary sources of wastewater in the brewery. Values for BOD and suspended solids can vary widely depending on the volume of water usage.

Operation	Source	Wastewater characteristics	Relative volume (% of total)
Brewhouse	Washing-rinsing	Cellulose, grain solids, hop residues, sugars, proteins, polyphenols, biochecmical oxygen demand (BOD) to 3,000 mg/L	25
Fermentation and lagering	Washing-rinsing	Yeast solids (6,000 mg/L), BOD up to 100,000 mg/L	17
Packaging	Filtration, kegging, bottling	Yeast solids and high BOD from beer and proteins (up to 135,000 mg/L), high pH from caustic cleaners	38
Utilities and miscellaneous	Beer spills, steam condensate, general cleaning	Relatively low solids and BOD (1,000 mg/L) but pH dependent on chemicals used	20

USG per day of discharge or at least 5% of the volumetric flow or BOD load to the municipal treatment facility. At six gallons of water per gallon of beer, this corresponds to a brewery output of about 25,000 bbls of beer per year. Even if the brewery is not required to pretreat, it may be economical to do so, depending on the sewerage taxes or fees and space availability at the brewery. In the United States of America, pretreatment usually makes economic sense only for relatively large breweries, producing between 150,000 and 300,000 bbls per year and wastewater generated at 2–4 times the beer production. If implemented, pretreatment could include equipment for equalization, pH adjustment, and BOD and/or TSS reduction by a combination of aerobic and anaerobic biological processes. Anaerobic processes have the advantage that methane gas is a byproduct of the treatment, a potential source of heat in the brewery.

7.7 Water and Wastewater Analyses

For the brewer, having a knowledge of the inorganic constituents in the water supply is key in understanding the chemistry of the mash, boil, fermentation and ultimately many of the subjective characteristics of the final product. Likewise, the level of organic material present in the effluent from the brewery will impact the load on the municipal treatment system and the associated fees assessed to the brewery.

Hardness

As described in Section 7.3, hardness is one of the most important characteristics of process water, impacting the mashing and boiling processes and ultimately the final product. Temporary hardness is also referred to as "carbonate hardness" and is the sum of the calcium and magnesium concentrations, both expressed as calcium carbonate in units of mg/L. There are two methods for quantifying hardness, the most accurate being a calculation based on the direct determination of calcium and magnesium ion concentrations:

$$\text{Hardness (mg/L of CaCO}_3 \text{ equivalent)} = 2.497[\text{Ca}] + 4.118[\text{Mg}]$$

where the ion concentrations in mg/L are determined using either atomic absorption spectrophotometry or the inductively coupled plasma method. The alternative method for determining carbonate hardness is based on EDTA titration. Ethylene diaminetetra acetic acid (EDTA) is a chelator that forms a soluble complex with certain metal cations. The method uses a colorimetric titration in which a dye, such as Eriochrome Black T or Calmagite, is first added to the water sample adjusted to a pH of 10.0. The solution will turn a wine-red color based on a reaction with the calcium and magnesium ions. EDTA is titrated into the solution, complexing the calcium and magnesium ions. Once all the ions have been complexed, the solution turns color from wine-red to blue indicating the endpoint. The titration must be performed at the correct pH of 10.0 and should be performed within 5 minutes to minimize precipitation of $CaCO_3$. The level of hardness, as mg/L of $CaCO_3$, is calculated based on the volume of EDTA titrant solution required to reach the endpoint.

Metal Cations

The concentrations of metal cations in solution is determined using either atomic absorption spectrometry or the inductively coupled plasma method. For the majority of cations of interest to the brewer, flame atomic absorption method can be used in which the samples are atomized by direct aspiration into an air-acetylene flame. A light source composed of the element in question is directed through the flame, a filter and onto a detector. Because the light emits at a wavelength characteristic to the element, the amount of absorption of the light by the atomized sample is proportional to the concentration of that particular element in the sample. An alternative method is based on inductively coupled plasma (ICP) emission spectrophotometry. The advantages to the ICP method include low detection limits for many elements, broad ranges in concentration (four to six orders of magnitude) and simultaneous multielement determinations. This is made possible by the extremely high temperatures generated in the plasma stream, 6000–8000 °K, resulting in complete dissociation of the molecules in the sample and efficient excitation to produce the ionic emission spectra. The emission spectrum is analyzed for intensity at different wavelengths, by a detector equipped with either a monochromator, a single photomultiplier and a sequential scanning mechanism or a polychromator with multiple exit slits and photomultipliers that provide simultaneous wavelength analysis. For most of the cations of interest to the brewer, sensitivity is as low as 0.01–0.03 mg/L up to concentrations as high as 100 mg/L without need for sample dilution.

Inorganic Anions

Analysis of the process water for individual anions, such as nitrate, phosphate and sulfate, can be accomplished with conventional colorimetric, electrometric or titrimetric methods. However, the use of ion chromatography is recommended based on several inherent advantages, including the elimination of hazardous reagents, small sample sizes, the rapidity of the analysis, ability to differentiate between the various halides (Br^-, Cl^- and F^-) and also the state of oxygenation of sulfur and nitrogen (for example SO_3^{2-} and SO_4^{2-}). The principle of the technique relies on the use of a series of anion exchange columns in which the anions are separated based on their relative affinities to the column matrix. The separation occurs under basic conditions in a carbonate-bicarbonate eluent. After separation, the anions flow through a hollow-fiber cation exchanger or micromembrane bathed in a strong acid solution. In this stage the anions are converted to their highly conductive acid forms, while the carbonate-bicarbonate eluent is converted to carbonic acid, which is only weakly conductive. The acidified anions are detected based on their conductivity and identified from the relative retention times. Quantification of the concentration is from peak height or area in comparison to standards. Typically, the minimum detectable concentrations are 0.1 mg/L but high levels of organic acids or individual anions may cause interference with other anions. It is also important to prefilter samples to remove any particles larger than 0.2 microns that may clog the columns.

Biochemical Oxygen Demand (BOD)

The concentration of biodegradable organic material in the wastewater is most often determined indirectly by performing a standardized test for biochemical oxygen demand. The test measures the quantity of oxygen consumed by bacteria as the waste is degraded while incubated at 20 °C, typically over a 5-day period as designated by the subscript on BOD_5. Unless an inhibiting chemical is used to prevent oxidation of nitrogenous compounds, the BOD_5 value will reflect the demand of both the carbon and nitrogen compounds in the wastewater. However, it may be desirable to separate the effects of the nitrogenous oxygen demand by measuring the organic nitrogen and ammonia content separately as described below. Nitrification can be eliminated during the BOD test by the addition of 2-chloro-6-(trichloro methyl) pyridine (TCMP) to the samples prior to the start of the test.

Key factors in obtaining accurate, reproducible values of BOD_5 are using appropriate dilution factors for the wastewater samples and ensuring a consistent seed culture of bacteria capable of performing the aerobic waste degradation. Dilution factors should be back-calculated based on estimated BOD_5 values. For best accuracy, after the 5-day incubation period there should be at least 1 mg/L of residual dissolved oxygen and a consumption of at least 2 mg/L. For example, if the expected BOD_5 value is between 1,000 and 5,000 mg/L and the initial value of dissolved oxygen (DO) is 9 mg/L at 20 °C, then the maximum change in DO should be 8 mg/L corresponding to 5,000 mg/L of BOD_5. So, the ideal dilution factor can be calculated as $8/5,000 = 0.16\%$. However, because this is an estimate of an unknown value, best practice would be to bracket this value and prepare several sample dilutions between 0.1% and 1.0%. Use of a seed culture is only necessary if the wastewater does not naturally contain a sufficient population of active microbes. This may be the case with high strength industrial wastewater as can be generated in some brewing operations. The best microbes to use are populations that have been adapted to the particular characteristics of the wastewater. Such a population can be generated in the laboratory by aerating a sample of the waste for several days until the microbes are established. Alternatively, a culture can be obtained from a local treatment facility or by purchasing a commercial seed preparation.

The specialized apparatus needed for BOD testing includes multiple bottles with 250 or 300 mL capacity, an incubator water bath set to 20 ± 1 °C, a dissolved oxygen probe designed for insertion into the bottles with an accuracy of ± 0.1 mg/L of DO. A check on the accuracy of the laboratory's procedure should be performed periodically using a standard test solution consisting of glucose plus glutamic acid, both at 150 mg/L. The average BOD_5 for this mixture should be about 200 mg/L with interlaboratory studies showing a standard deviation of 30 mg/L.

Organic Nitrogen

Brewer's wort contains significant quantities of organic nitrogen in the form of proteins, peptides and amino acids or free-amino nitrogen (FAN). Most of the FAN is consumed by the yeast during fermentation, however, residual FAN from washing operations in the brewhouse plus peptides and proteins associated with the trub and spent yeast will be part of the wastewater effluent and may contribute significantly to the biochemical oxygen demand. It may be desirable to separate the contributions of organic nitrogen from carbonaceous sources of BOD. The standard for determining organic nitrogen is called the *Kjeldahl method* and is the sum of organic nitrogen and ammonia nitrogen. The principle of the Kjeldahl method is based on the conversion of organic nitrogen into ammonium ions. This is accomplished by digestion of the sample in special Kjeldahl flasks at a temperature between 375 and 385 °C in the presence of sulfuric acid, potassium sulfate and copper sulfate catalyst. When the conversion is complete, base is added to produce alkaline conditions and the resulting ammonia is distilled off and absorbed in a boric or sulfuric acid solution. The ammonia concentration is then determined either by titration or with an ammonia-selective electrode (see below). The end result is the sum of total organic nitrogen and ammonia that was initially present in the sample.

Ammonia Nitrogen

Brewer's wort can have small quantities of ammonia nitrogen, which is largely consumed by the yeast during fermentation, but can also make its way into the brewery effluent from cleaning operations in the brewhouse. Ammonia is a readily assimilable form of nitrogen so, most municipalities have limits on the acceptable level of ammonia that can be directly discharged without pretreatment, usually between 30–40 mg/L. Currently the most common method for analysis of ammonia in water is with the use of an ammonia-selective electrode. These electrodes are accurate over a wide range of concentrations (0.03–1400 mg/L), are easy and fast to use and unlike other methods, do not require a preliminary distillation step unless total organic nitrogen is to be measured (see Kjeldahl method). The electrode functions based on the diffusion of ammonia through a gas-permeable membrane into an internal solution of ammonium chloride. Before analysis, the alkalinity of the sample is adjusted with a strong base to a pH above 11, thereby converting all NH_4^+ ions into dissolved ammonia gas. The NH_3 diffuses through the membrane affecting the pH of the internal solution, which is measured by the internal pH electrodes in proportion to the ammonia concentration in the sample. Generally, the method is free from interferences, although the presence of some amines can increase the reading.

Total and Suspended Solids

Considerable quantities of solid materials are disposed off in brewery effluent, being generated from spent grain residue, hop residues, trub and spent yeast. Depending on the quantity, these suspended solids can pose problems for the municipal treatment system and the brewery may be faced with extra surcharges. It is important to understand the terminology for the solids fraction of wastewater. "Total solids" refers to the amount of solid residue left over from a sample after drying to constant weight and will include both total suspended solids and total dissolved solids. Although "suspended solids" can be determined differently depending on the pore size of the filtration media, "dissolved solids" refers to the solids that will pass through an absolute filter with pore size of 2.0 micron or less. "Volatile solids" refers to the weight loss of solids from incineration and is often considered as a measure of the organic content of the solids, although this is not strictly true as some mineral salts also undergo decomposition at high temperature.

The drying of a sample to determine total solids is usually performed at a temperature between 103 and 105 °C. The sample is placed in a weighed dish and maintained at the drying temperature in an oven or incubator until constant weight is achieved. The weight of residue is divided by the sample weight to obtain the fraction of total solids, or by the sample volume to determine concentration. A sample volume is chosen such that a residue weight of at least 10 mg is obtained and samples are tested at least in duplicate.

The fraction of dissolved solids in either process water or wastewater is obtained by filtering the sample through a standard glass fiber filter. Note that if wastewater samples contains high amounts of suspended solids, the samples may need coarse pre-filtering to prevent clogging of the filter. The collected filtrate is placed in a pre-weighed dish and dried at 180 °C in a drying oven. Samples must be cooled in a desiccator before weighing and then placed back in the oven until constant weight or until there is less than a 0.5 mg change. Sample volume should be sufficient to provide at least 10 mg of dissolved solids residue.

To determine the volatile fraction of residue collected in either of the above analyses, the residue is ignited at a temperature of 550 °C to constant weight, although, typically 15–20 minutes should suffice for up to 200 mg of residue. The weight loss from ignition is defined as the volatile solids content and is an approximation of the organic fraction in the residue. It should be noted that some volatilization of organic matter can occur while oven drying at 105 °C, so, for samples with a high content of organic solids, drying temperatures are sometimes reduced.

7.8 Water Management and Conservation

For a brewery to implement an effective water management strategy, the first step is to collect pertinent data for each step in the brewing process:

1. Volumes of water usage
2. Volumes of water discharged to waste
3. The characteristics of the discharged water
4. Unnecessary water usage and opportunities for reduction

Volumetric data is best collected with the use of submeters, connected to dedicated pipes that supply water only to stations associated with specific operations. In this way a pattern of water usage can be established over time for each different brewery operation so that data becomes statistically relevant. If submeters are not available, then water consumption can be estimated based on a measured volumetric flow rate and a recorded time of usage for a given operation, although this method is tedious and far less accurate. During each operation, records should also be kept of the fraction of water used as process water versus the fraction that is discharged to waste and, if possible, samples obtained of the wastewater for analysis. The municipality may be obtaining 24-hour composite wastewater samples from the entire brewery discharge, but this data would not be very helpful in characterizing contributions from specific operations. Once the data has been reviewed for consistency and accuracy, and Key Performance Indicators (KPIs) defined, the data can be considered as the "benchmark" values against which the effects of changes in the water management strategy will be compared. Typical KPIs may include: total water volume per month, volume of wastewater generated per month and volume of water for a specific use (for example, utility cooling water). However, because water is closely related to the level of activity in the brewery, the KPIs are best if they reflect the ratio between water usage and volume of beer produced. A reduction strategy could start by developing a timetable to lower the KPIs to realistic targets, while in consideration of other priorities within the brewery.

Some strategies for reducing the water/beer ratio can be relatively simple, for example by fixing leaking pipes and valves, while some are complex and much more costly, such as treating specific wastewater streams for closed loop recycle. Often the most cost effective measures are not actually focused on decreased water use but instead on increasing yield in brewing operations and reducing beer losses. In general, the methods for reducing water use are based on:

1. Monitoring and adjusting water flow rates for specific tasks
2. Modifying or replacing equipment to increase efficiency
3. Reusing or recycling water when possible
4. Modifying or replacing a process step

Often CIP systems have not been optimized for specific vessels or specific residues and a single cleaning routine is used in all circumstances. Significant savings in both water and chemical use can be realized by performing rigorous cleaning validation for each different vessel for each specific recipe, adjusting the cleaning routine as appropriate so as not to waste time and resources. For example, when cleaning the fermenters, top versus bottom fermentations can pose different challenges for the CIP system and the cleaning cycles should be adjusted accordingly.

7.9 Topics for Discussion

- What considerations are there likely to be when deciding whether to obtain the brewery water from a municipal supply versus a dedicated well?
- How can a brewery produce both high quality ales and lagers when using the same source of process water?
- Describe some of the limitations of the BOD_5 test for characterizing the organic wastewater load.
- What activities would need to be performed in order to validate a clean-in-place process for a specific vessel in the brewery?
- List some strategies that could be considered when implementing a water reduction program.

7.10 Further Reading

1. De Clerk JA. *Textbook of Brewing*. Volume 1. Chapman & Hall Ltd; 1957.
2. In: Eaton AD, Clesceri LS, Greenberg AE, ed. *Standard Methods for the Examination of Water and Wastewater*. 19th ed. American Public Health Association; 1995.
3. *The Practical Brewer*. 3th ed. Wauwatosa, Wisconsin, USA: Master Brewers Association of the Americas; 1999.
4. *Water and Wastewater: Treatment/Volume Reduction Manual*. Boulder Colorado, USA: Brewers Association; 2017.

Chapter 8

Beer Chemistry and Testing

8.1 Introduction

Brewing beer is the art of steeping a starch source (e.g., barley malt) in water, possibly with the addition of flavor herbs like hops, and subsequent fermentation with bacteria or yeast. This process has been around for thousands of years and most brewing was done by non-scientists. Process control was mostly limited to following closely guarded recipes and procedures. The principles of brewing science as we know it today were for the most part not understood and the variations from one batch to the next must have been huge. Random infections with bacteria (e.g., *Lactobacillus*) or wild yeast must have happened but were not well understood and were treated as a mystery.

The science of brewing became much better understood in the 19th century, when Louis Pasteur discovered the important role yeast played in fermentation, and the German chemists Justus von Liebig and Friedrich Wöhler started to understand the important roles of "organic" molecules. In places where beer was brewed, brewing research laboratories and fermentation institutes were founded and over time developed into world renowned teaching and research institutions. Next to the oldest brewery in the world in Weihenstephan, close to Munich, Germany, the Weihenstephan Brauereihochschule was founded in 1865. The Dr. Siebel Analytical Laboratory in Chicago was founded only three years later in 1868.

Today, the American Society of Brewing Chemists (ASBC), the Master Brewers Association of the Americas (MBAA), the European Brewery Convention (EBC), the Brewery Convention of Japan (BCOJ) and the Institute of Brewing and Distilling (IBD) all play an integral part in developing methods for modern quality control procedures in brewing, malting, hop and finished beverage testing laboratories throughout the world. They all share the commitment to create and exchange knowledge in fermentation science, bioprocessing and biotechnology.

Is it possible to brew a high-quality beverage without an in-depth understanding of fermentation science and a fully equipped testing laboratory? Of course it is, just like a good chef can craft awesome meals day after day. However, a good understanding of brewing science and bioprocessing combined with a thorough quality testing program can help the commercial brewer as well as the homebrewer to brew a high quality, consistent, shelf-stable product.

The qualitative and quantitative testing of the finished product occurs according to the following three main criteria:

- Taste, aroma and appearance of the beer (olfactory evaluation)
- Microbiological testing for spoilage organisms
- Chemical/physical evaluation (analytical testing)

8.2 Taste Testing

While many ingredients and properties of the finished beer (e.g., amount and stability of the foam, color, turbidity, alcohol content) can be accurately measured with analytical testing methods, the purity and crispness of the taste and aroma (smell), the fine differences in bitterness and flavor profiles, smoothness of mouthfeel, these are factors that cannot be captured by analytical methods. But these are exactly the factors that the consumers are primarily interested in. Once consumers have selected their favorite beer, they would like to have the same consistent experience over and over again.

Thus, taste testing is one of the three cornerstones of quality control of the finished product. Taste testing can serve two distinctly different purposes. A test to evaluate how much certain consumer groups "like" a beer is usually done externally with current and potential customers, who for the most part are untrained beer enthusiasts. These tests are very valuable for marketing and market research purposes, to find out what customers like and what finished beer attributes they are actually willing to pay for. Alternatively, the brewer can submit beers to beer judging events, were beers are rated and ranked by a trained panel of judges (e.g., accredited by the Beer Judge Certification Program (BJCP) or the Cicerone Certification Program) according to different style categories and judging guidelines. The taste testing for internal quality control purposes is usually done by a diverse group of internal personnel. This group of tasters has to be well trained and calibrated, and has to have a very refined ability to distinguish between fine differences in flavor and aroma. An initial evaluation of a taster's ability to differentiate between beers, one can test them for example with a beer that has been diluted with 10% deionized water (dilution comparison), a beer to which 4 g/L of sugar has been added (sweetness comparison), or a beer to which 4 mg iso-α-humulone acid has been added (bitterness comparison). The samples to be compared are stored at a cool temperature and then served "covered" to the tasters by the tester in neutral small glasses. Covered means that the individual glasses are coded in a way that it does not give the taster any hint of the origin or altercation of the beer. The test can be conducted as a Duo-Test, in which the differences of two samples (AB) have to be detected and explained, or a Triangle-Test, in which the taster receives one of the two samples twice, in various orders (e.g., AAB, BBA, ABA, ABB, etc.). The taster must determine the one sample that is different and accurately explain the difference. The tasting panel can thus be trained and calibrated to different sensitivity levels with various flavor and aroma standards, including off-flavors. In addition to the comparative testing the taster panel can also evaluate a beer by a point scale for different characteristics like aroma, purity and crispness of the taste, mouthfeel, lingering taste, bitterness quality and more. Eventually, the trained and calibrated tasting panel can evaluate the beer quality from batch to batch, after storage under various conditions (to evaluate shelf life stability), new beer styles and recipes, process troubleshooting and samples returned by unhappy customers.

8.3 Microbiological Testing

Most modern-day breweries keep a very clean, food grade production environment. Many breweries embrace the internationally recognized HACCP system for reducing the risk of safety hazards in food (and beverage) production. HACCP stands for Hazard Analysis and Critical Control Points. A HACCP system requires that potential hazards are identified and controlled at specific points in the process. Despite all the efforts of keeping a clean production environment, it is possible that microorganisms make it into the finished packaged product. Once in the final package, certain organisms can grow and multiply, form a layer at the bottom of the package (can, bottle, keg), create haze in the beer, and through the formation of certain metabolic byproducts can change the flavor and aroma of the beer, spoil it, or render it undrinkable or in rare cases, make people sick. Thus, it is imperative to discover the presence of such organisms as early as possible, find out how they entered the process chain, and take measures to stop their growth and entry into the finished beer. Not all microorganisms that are found in the finished beer are necessarily bad or spoilage organisms. Basically, their negative impact can be classified according to three categories:

- The harmless microbiome
- The potential beer spoilers
- The obligatory beer spoilers

The harmless microbiome can consist of fungal spores, left over brewers' yeast from the fermentation process, harmless bacteria and yeasts. Most of these cannot thrive in the finished beer due to the absence of oxygen, the presence of a low pH (high acidity), alcohol, and the terpenes and humulones from the hops, which can have antiseptic properties, even at low concentrations. Some breweries add a small amount of beer yeast during or just prior to the bottling process, in order to consume quickly any remaining oxygen that may still be present in the final package after packaging (The Total Package Oxygen, TPO).

The potential beer spoilage organisms can only multiply in beer when certain favorable conditions are present in the beer: high oxygen content, high pH (4.7 or higher), low hops bitterness or lower alcohol content. These potential spoilage organisms can occur: e.g., *Enterobacteriaceae, Streptococcus lactis, Lactobacillus casei.* Some of these microbes can produce lactic acid and other undesirable off-flavor byproducts. Certain yeasts can also contribute to beer spoilage and instability. These so-called wild yeasts include *Saccharomyces diastaticus, and Saccharomyces ellipsoideus,* but also *Brettanomyces, Torulopsis, Hansenula,* and *Candida.* These wild yeasts often cause haze, make nutrients unavailable for the culture yeast, and produce undesirable off-flavors.

The obligatory beer spoilage organisms are of a major concern to the breweries. These include the anaerobic bacteria like *Pectinatus cerevisiiphilus* and *Megasphaera cerevisiae.* The brewer actually creates a favorable environment for these spoilage organisms as they thrive in low oxygen and low pH environments. These organisms form certain metabolites that lead to very unpleasant tasting and smelling beer, rendering it undrinkable. They can hide dormant in the biofilms that are created by the ubiquitous acetic acid forming bacteria, and become active when conditions become favorable for them.

Among the obligatory spoilage organisms is also a group of aerobic microorganisms, which thrive in the presence of oxygen (but some of them can also live under anaerobic conditions). These microbes can produce significant amounts of acetic acid, by oxidizing the alcohol and rendering the beer sour. Among these microbes are: *Pediococcus damnosus, Lactobacillus brevis, Lactobacillus Lindneri,* and *Lactobacillus frigidus.* Spoilage organisms can hide in the brewhouse where the Clean-in-place (CIP) cleaning does not reach adequately, e.g., in hoses and transfer lines, valves, seals, pumps, meters, sampling ports, package rinse process, bottle capper, can sealer, filler ports, etc.

Testing for these spoilage organisms is typically done in a conventional microbiology laboratory, using customized agar plates in petri dishes, on which the beer sample gets plated out under aseptic conditions. The plates are then incubated under various conditions (aerobic, anaerobic, temperature) and evaluated by a trained scientist. Incubation times can vary from 24–48 hours and up to 5 days for certain slow growing organisms. This time frame may be too long as a waiting and holding period before product release into the market. A faster, though more expensive method is based on quantitative Polymerase Chain Reaction (qPCR) analysis, in which the DNA fingerprint of the spoilage organisms can be detected quantitatively and differentiated for each spoiler species in only a few hours' time. Disadvantages of this method include the high cost for science trained personnel, laboratory facility, equipment and reagents, but also the potential for false positive results, as this technique will also detect DNA fragments in dead organisms, which may not have survived the CIP process but their DNA is still present in the finished product.

A quick test for the presence/absence of live microorganisms can be done quickly and conveniently with ATP swabs, which then get analyzed in a handheld ATP-Luminometer. The measurement takes only a minute and can quickly determine if CIP and other cleaning processes have been successful and adequate.

Other microbiological tests evaluate the brewing water, wash water, yeast purity, health (vitality), and viability, flocculation properties of yeast, air, surfaces, equipment and more. Cleanliness and verification of successful cleaning procedures is of utmost importance. Refer to Chapter 7 for additional information on this topic.

Last but not least, most breweries keep stability samples from each batch. These samples get stored under controlled conditions, often times in a cooler or refrigerator. These samples get examined at the end of the anticipated shelf life period to see if haze, bottom settlement, or off-flavors, aromas or other visible changes have occurred. They typically also undergo a tasting and a microbiological examination, as well as certain chemical storage stability tests. These tests provide assurance that the finished product is shelf-stable for the anticipated period of time and within the quality specification that the brewery defines for each beer.

8.4 Chemical Analyses

While many flaws, problems and potential spoilage issues can be detected by visual, olfactory (smell and taste) and microbiological evaluation, certain important parameters can only be detected and measured by chemical analysis or physical parameter testing. This testing can either be done right at the brewery in the brew house, or a dedicated laboratory in the brewery. However, some of the tests require very sophisticated analytical instrumentation that is expensive to buy, maintain and operate and is likely beyond the financial means of most breweries and is usually offered by third party laboratories.

In order to produce a beer of consistent quality, multiple parameters need to be continuously measured and controlled, batch by batch. These parameters are tested throughout different stages of the brewing process, starting with certain parameters for the ingredients, all the way to the finished, packaged beer. Based on these parameters, the brew master then can establish a sheet of detailed specifications for each beer and production stage. Sophisticated breweries track more than 50 parameters for each production batch with very tightly defined tolerance brackets for each parameter. One of the most important parameters in beer is the alcohol content.

Alcohol and Specific Gravity

The alcohol content of beer is important in order to be compliant with the rules of the U.S. Department of the Treasury, Alcohol and Tobacco Tax and Trade Bureau, abbreviated TTB. While beer in the United States is not taxed based on alcohol content, the TTB regulates the beer labels (and recipes). Currently (27CFR A, part 7, C), the TTB does not require that alcoholic content is stated on the label, although some states do require this. When a state requires that alcohol content be shown on the label, then it has to be stated either in percentage of alcohol by weight (% ABW) or volume (% ABV). The two values can be distinctly different based on the final gravity of the finished beer. Some states regulate a maximum alcohol limit of beer, e.g., in the State of North Carolina the maximum allowed alcohol content in beer is capped at 15% ABV. The terms "low alcohol" or "reduced alcohol" may be used only on malt beverage products containing less than 2.5% ABV. The term "nonalcoholic" may be used on malt beverage products, provided the statement contains "less than 0.5% ABV" appears in direct conjunction with it. In general in the US, any beverage containing less than 0.5% ABV is regulated as a nonalcoholic beverage (e.g., Kombucha). The term "alcohol free" may be used only on malt beverage products that contain no alcohol (0.0% ABV). The TTB allows a tolerance of 0.3% below or above the stated % ABV amount, except when it comes to nonalcoholic beer or alcohol free beer, in which case these tolerances do not apply.

Accurate measurement of the alcohol in beer is also important for the brewer to see how efficient and consistent the brewing process was for a particular batch of beer. Because of batch to batch process variability due to changes in ingredients, temperature in the brewhouse and yeast vitality (and other influencing factors), it is not possible to accurately predict the amount of alcohol in the finished beer. The actual alcohol contents can fluctuate between ± 0.3% ABV or more. There are several different ways to measure alcohol contents in the finished beer.

This can be done by measuring the Original Gravity (OG) of the wort prior to fermentation and the Final Gravity (FG) after the fermentation has been completed. The conversion formula is pretty simple:

$$\% \text{ ABV} = (\text{OG-FG}) \times 131.25$$

So, using this formula with a beer having an OG of 1.045 and a FG of 1.008, the alcohol content in the finished beer would be 4.86% ABV. The gravity can be measured with a simple hydrometer or saccharometer, a graduated glass spindle that sinks deeper into the liquid the less dense the liquid is. The more molecules, in particular fermentable sugars, have been extracted into the wort, the higher the OG. The gravity (or density, both terms are used synonymously in this text) measurement is temperature sensitive and needs to be temperature corrected. A more accurate density meter is based on an

Figure 8.1. An automated bench-top instrument from Anton Paar for accurately measuring specific gravity, or °Plato, using an oscillating U-tube.

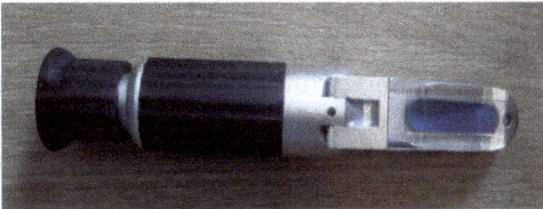

Figure 8.2. A portable refractometer used for measuring sugar concentration in liquid samples.

oscillating glass U-tube with built-in temperature correction as used by the Anton Paar instrument pictured in Figure 8.1. It is important to note that the alcohol measurement via density is only an approximation. It works quite well for medium strength beers that do not have too many unfermentable dissolved solids in the original extract. This is because the components in the extract (dissolved solids) increase the density, while ethanol decreases the density. Often, a quick determination of density can be obtained with a refractometer (see Figure 8.2), which measures the refraction of light as it passes through a liquid layer at an angle. While inexpensive, light and portable, these instruments also just give a rough estimation of density. A better approximation can be achieved by a significantly more expensive instrument that combines gravity, temperature and a

Near-Infrared (NIR) absorbance spectrophotometer. However, these instruments are not accurate enough for very high gravity beers (Belgian Triples, Barley Wine, high alcohol stouts etc.), for which an alcohol distillation step is required, or very low alcohol beverages (e.g., non-alcohol beer, alcohol free beer, kombucha). The most accurate analytical method to determine alcohol contents in any beverage is based on gas chromatography in conjunction with a mass spectrometer detector and isotope labelled internal reference standard.

The Original Gravity (OG) measurement of the original extract (or apparent extract) in the finished (boiled) wort and the measurement of the Final Gravity (FG) of the finished beer are still very valuable parameters for the brew master. They were actually some of the very first measurements that were introduced into commercial breweries after the introduction of temperature measurements. The OG give an indication of the strength of the original extract, or apparent extract. It is usually measured in the wort after the boil, just prior to fermentation. As beer is being fermented — or attenuated — its gravity decreases, because the heavier sugars are being fermented to the lighter alcohol, which stays mostly in the beer, and to carbon dioxide gas, which mostly escapes from the fermentation vessel. Thus, attenuation is a measurement of the extent to which sugars have been converted to alcohol. And, if the actual alcohol amount is known in the finished beer, the brew master can calculate back to determine how much fermentable sugar must have been present prior to fermentation. This is often called the Real Extract (RE). It represents the corrected attenuation, as it corrects for the fact that ethanol has only 79% of the gravity of pure water. For the calculation of RE from OG and FG, two correction factors (0.1808 and 0.8192, respectively) are being used to correct for the presence of alcohol in the finished beer and the absence of alcohol in the unfermented wort:

$$\text{Real Extract (RE)} = (0.1808 \times \text{OG}) + (0.8192 \times \text{FG})$$

If we take the OG and FG values from our earlier example the RE is:

$$\text{RE} = (0.1808 \times 1.045) + (0.8192 \times 1.008) = 1.015$$

The difference between the apparent extract and the real extract yields the residual extract, which mostly contains the unfermentable sugars like lactose, larger dextrins and oligosaccharides, but also leftover fermented sugars that just did not get converted to alcohol by the yeast in the fermentation process, possibly due to lower yeast vitality or other fermentation parameters.

$$\text{Residual Extract} = \text{OG} - \text{RE}$$

In our example:

$$\text{Residual Extract} = 1.045 - 1.015 = 0.030$$

Another way of looking at our example is as follows: 1 liter of water has a density of 1.000 and thus weighs 1,000g. A liter of wort with an OG of 1.045 contains 45g original extract. Of these 45g, 30g constitute the residual extract that is left over after fermentation, and 15g fermentable sugars that were converted to alcohol in the finished beer.

Attenuation, or how well the fermentation converted the original extract to the finished beer, is often presented as the difference of the Original Gravity of the unfermented wort and the Final Gravity of the finished beer, expressed as a percentage of the Original Gravity with the following formula:

$$\text{Attenuation \%} = (\text{OG-FG}) / (\text{OG} - 1) \times 100$$

In our example:

$$\text{Attenuation \%} = (1.045 - 1.008) / (1.045 - 1) \times 100 = 82\%$$

The attenuation gives the brew master a picture of the brewhouse efficiency, yeast and malt efficiency and overall efficiency of the brew process in a particular brewery, for a particular beer recipe. It is an important parameter for the brewer to measure in every batch in order to detect abnormalities early and to ensure batch to batch consistency.

Acidity

Acidity measurements are important in brewing, and are typically measured as pH value. pH values can range from 1–14, where the range from 1–7 is called acidic, 7 is neutral (distilled water), and 7–14 is called basic or alkaline. 1 is the highest acidic level, and 6.9 is the lowest. The pH measurement is temperature dependent, and most pH meters use a combination glass electrode with automatic temperature correction (ATC). Today, inexpensive handheld pH meters are readily available and can quickly be calibrated in the brewery with calibration buffers. Typical pH values range from 5.0–5.6 in a finished wort (5.0–5.2 is preferred), and in most finished beers from 4.2–4.6, except for some sour beers that can have pH values between 3.0 and 3.6. The pH is very important as many enzymatic processes but also most microbial processes and growth are heavily dependent on a narrow pH range for their optimum conditions.

The pH should be tracked throughout the fermentation process on a daily basis, as the yeast forms acids as metabolic byproducts and releases them into the beer, lowering the pH. The more vigorous the fermentation progresses, the faster the pH drops. So, as the pH changes from day to day, the brewer can tell how well the fermentation is progressing. Acidity in the finished beer is also a very important flavor aspect and suppresses microbial growth.

Color

Even though it does not sound that important, for many beers, color is a significant part of their characteristic appearance and customer appeal as presented in Figure 8.3. Getting the same color every time is not an easy task, especially when the ingredients vary from batch to batch. The easiest and most traditional way to measure color is with color cards or charts, e.g., with a device called a Lovibond tintometer, that utilizes different colored disks for comparison. It turns out that the comparison of a liquid beer to a color chart is very challenging and error prone.

Today, many beverage laboratories measure color with a spectrophotometer (see Figure 8.4). Different methods are described by the various organizations, e.g., ASBC, EBC. Some utilize the absorbance of visible light at a single wavelength (e.g., 430 nm), others have a more complex approach by comparing absorption at different

Figure 8.3. Three distinctly different beer styles and associated colors: Pilsner lager, American Pale Ale and a Dark Porter.

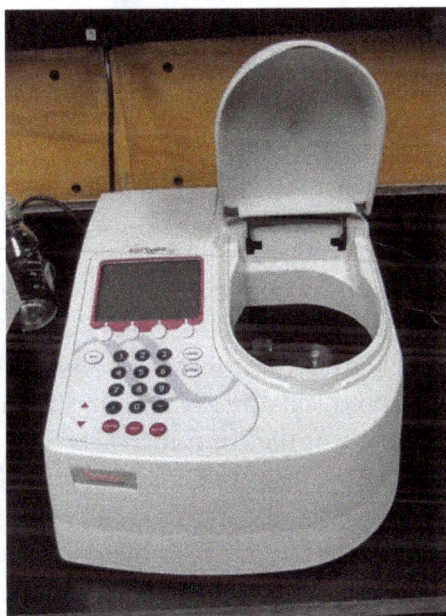

Figure 8.4. A variable wavelength spectrophotometer for measuring color and bitterness.

wavelengths and a calculation formula, e.g., the tristimulus method, to determine relative lightness, darkness, and hue. The Standard Reference Method determines color in the "SRM" unit via spectro-photometer. It is a useful, standardized parameter, and less error prone than doing a visual comparison of the beer to color charts, as light intensity, spectrum, angle, size, shape and material of the glass etc. can make a difference. However, it has been shown that beers with the same SRM determined by a spectrophotometer still had color differences that were detectable by the eyes of well-trained judges. Exact measurements with a spectrophotometer are particularly challenging for beers with very dark colors or high degree of turbidity.

Turbidity

Turbidity, or clarity and brightness is another important appearance parameter in beer. The lack of turbidity and pronounced clarity and brightness in European lager beers and traditional American light lagers has been a highly regarded quality parameter for many years. Certain other beer styles, in particular the German Hefeweizen, traditionally have been hazy by design and style, and distinguish themselves from their sister beer, the perfectly clear Kristallweizen "Crystal Wheat," which has essentially no turbidity and a significantly different flavor profile. Innovative craft beers in North America appear to subvert the existing paradigm of high clarity beers as a high-quality trait.

In particular the advancements of New England Imperial Pale Ales, as pictured in Figure 8.5, show a very high turbidity as part of their signature taste and quality profile and thus created a new, desirable beer style in the eyes of the consumer — a beer fault of the past turned into a new signature feature. On a worldwide basis, with a few exceptions as noted above, clarity of beer is still a very important quality factor. In particular chill haze, which is caused by certain ingredients in beer (polysaccharides and polyphenols) is still considered a problem by many brewers and perceived as a flaw by consumers. Turbidity is measured by turbidity meters, which measure light scattering (due to particles) at various angles. The more light is scattered, the more haze / turbidity is detected in the beer. Modern day portable turbidity meters can detect turbidity with more sensitivity than the human eye. This in turn creates a challenge for the brew master, because a subjective decision has to be made as to whether a particular beer passes the internal (and external) specifications or not. Turbidity

Figure 8.5.   A hazy American IPA style.

plays a very important role in the entire beverage industry, whether it be fruit juices, wine, distilled spirits or beer.

Bitterness

One of the most distinctive characteristics of beer is the bitterness that both contrasts and compliments the other flavor notes, whether the bitterness be subtle or dominating, it is hard to imagine a beer without hops. As described in Chapter 6, there are many different varieties of hops that in addition to bitterness from the isomerized α-acids, can provide other flavor and aroma qualities to the beer, from spicy to floral and fruity. Yet, it is the perceived bitterness that often defines a style for the consumer and brewers will typically emphasize the level of bitterness in their beers as a selling point, especially with the American IPA style.

Either on the board in the taproom or on the can or bottle, bitterness is typically stated as International Bitterness Units (IBU). Some brewers estimate the IBU level of a particular beer with the use of models that predict bitterness based on the use of a certain quantity of hops with a specific α-acid content. These models can provide guidance for recipe design, however, because of the large number of process variables that can affect the final bitterness, they are not very accurate. It is, therefore, recommended that an actual bitterness test be performed on all beer made with a new recipe, new variety or batch of hops or if boiling conditions have been modified, as these factors can all affect the final bitterness.

The most practical and convenient method for most brewers to measure bitterness is by using the Beer-23 Method A as described by the American Society of Brewing Chemists (ASBC). The only lab equipment required is a UV-VIS spectrophotometer, a centrifuge (for 50 mL tubes) and a low speed mechanical shaker. The method describes the procedure for extracting the isomerized acids from the beer with a solvent and then determining bitterness units (BU) based on absorbance measured at a wavelength of 275 nm.

It is also possible to perform an alternative method in which the actual concentration of iso-α-acids in mg/L is determined by High Performance Liquid Chromatography (HPLC). This is described as Method C under the ASBC method Beer-23.

8.5 Dissolved Oxygen

To give the yeast a kick-start at the beginning of the fermentation, oxygen is usually added to the wort after it has been cooled and during the transfer from the kettle to the fermenter. The oxygen dissolved in the wort is quickly consumed by the yeast and from this point on, the fermentation should proceed under anaerobic conditions. Subsequently, to prolong beer freshness, any further exposure to air should be avoided. Although this may be impractical as exposure to air can occur during transfer operations through hoses or pipes, during filtration or centrifugation, in the bright tank while filling and during packaging.

Eventual staling of the beer is largely a result of reactions mediated by free oxygen radicals (or Reactive Oxygen Species — ROS) forming carbonyls in the beer such as trans-2-nonenal, which imparts a cardboard or paperlike taste and aroma. Fresh beer will have initial levels of trans-2-nonenal below the taste threshold of 50–100 ng/L (0.05–0.1 parts per billion), but then sometimes rising to more than 0.2 parts per billion (ppb) as the beer ages. Although eventually some oxidation will occur, best practices can limit exposure to air and provide for a reasonable storage life for the beer, especially under refrigerated conditions.

A good prediction of the storage life can be made by measuring the antioxidant potential of the beer using Electron Paramagnetic Resonance (EPR). This technique is based on the measurement of the rate at which free oxygen radicals are being formed by exposure of the beer sample to a strong microwave field. The length of the lag period during which there is no increase in the free radicals is indicative of the beer's antioxidant potential and has been correlated with the longevity of product freshness. The antioxidant potential of the beer is affected by the recipe and beer style, including hops and malt, the darker beers typically being higher in polyphenols. Additives with antioxidant properties, such as ascorbic acid and potassium bisulfite, can also extend storage life.

Another useful concept is "Total Package Oxygen" (TPO), which is the sum of the dissolved oxygen in the beer itself and the oxygen content, if any, in the headspace of the package, whether it be a can, bottle or keg. TPO should be as low as possible, but is typically in the range of 40–150 ppb, with a maximum of 10–60 ppb of oxygen dissolved in the beer. Determination of oxygen concentration requires the use of an accurate instrument, most often a probe that can be inserted directly into the package with a piercing device or an in-line detector that tests samples withdrawn through a tube from a tank or pipe. Figure 8.6 shows an example of such an instrument.

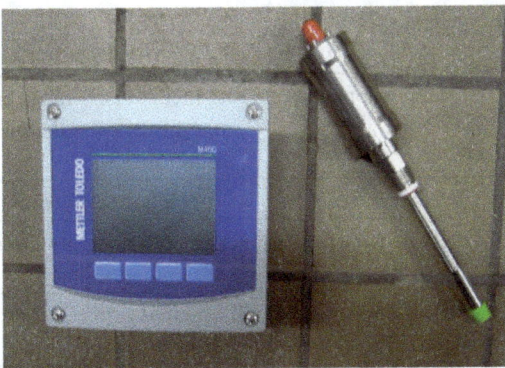

Figure 8.6. A highly sensitive dissolved oxygen probe and meter that employs a light-based sensor technology.

8.6 Undesirable Flavors and Aromas

It is difficult to generalize as to which of the up to 200 or more possible flavor and aroma compounds are desirable or instead considered as off-flavors in any particular beer. The decision is ultimately up to the subjective likes and dislikes of the drinker. Some flavors and aromas are considered as acceptable or even desired in some beer styles, while in other styles are considered as objectionable. Concentration also matters, as a subtle banana aroma from the presence of isoamyl acetate may be a positive attribute in certain lagers, but if it is too strong it becomes detrimental. The level at which a particular component is perceived (referred to as the taste threshold, as shown in Table 8.1) is also highly variable among individuals and can be significantly affected by the other components which comprise the complex beer matrix.

The accurate identification and quantification of the flavor and aroma components of beer is not straightforward. As many of the taste and aroma compounds are volatile, such as aldehydes, esters and alcohols, the most common technique for analysis is by headspace Gas

Table 8.1. Chemical components responsible for common taste or aroma defects in beer. Taste thresholds are approximate and vary by individual and the beer matrix. Data is based on information provided in The Defects Wheel for Beer by Susan Langstaff.

Chemical component	Perceived sensory effects	Possible source	Taste threshold
Acetaldehyde	Green apple, sherry, nutty	Yeast metabolism, microbial contamination, ethanol oxidation	5–15 mg/L
Acetic acid	Vinegar, sour	Microbial contamination,	130 mg/L
Diacetyl (2,3-butanedione)	Rancid butter, butterscotch	Yeast metabolism, microbial contamination	10–40 mg/L
2,6-dichlorophenol	Antiseptic, vinyl plastic	Chlorinated process water, sanitizers	5 mg/L
Dimethyl sulfide (DMS)	Canned corn, molasses	Raw materials, microbial metabolism	25 mg/L

Table 8.1.   (*Continued*)

Chemical component	Perceived sensory effects	Possible source	Taste threshold
Hydrogen sulfide	Rotten egg, sulfur	Microbial metabolism	4 mg/L
Isoamyl acetate	Banana, fruity	Yeast metabolism, especially at elevated temperature	1–2 mg/L
Isovaleric acid	Cheesy, putrid	Hop oxidation, bacterial contamination	1 mg/L
Mercaptan (3-methyl-2-bu-tene-1-thiol)	Skunky	Light exposure and iso-humulone reaction with riboflavin	4 ng/L
Trans-2-nonenal	Cardboard, paper	Exposure of beer to air, oxidation during storage	50–100 ng/L
4-vinyl guaiacol	Phenolic, cloves	Extracted phenols from malt, wild or spoilage yeast metabolism	0.2 mg/L

Chromatography (GC) using Flame Ionization Detection (FID) as described in the ASBC Method Beer-48. Identification of the various volatile components is based on column retention time while quantification requires frequent calibration of the instrument using known standards.

8.7 Foam Stability

The ability of the beer to hold a stable layer of foam at the top of the glass has traditionally been one of the most important indicators of quality. It is a sign that the beer has been made from high quality ingredients and is a reflection of the brewer's art (see Figure 8.7). It is only recently, with the rising popularity of drinking directly from cans and bottles, that the presence of foam has no longer been a consideration for many consumers. However, even if the visual appeal of the foam has lost some of its importance, the foam stability is a complex phenomenon and is still an indicator of many aspects of the underlying beer chemistry.

Figure 8.7.   Examples of beer with good head formation and foam stability after pouring.

Prior to packaging, the beer is supersaturated with CO_2 gas under pressure. If the pressure is released, bubbles will be produced as the dissolved carbon dioxide gas will tend toward equilibrium with the atmosphere. Once a bottle or a can is opened, this will occur slowly over time. However, most beer is opened for the purpose of drinking, so, soon after opening the beer is typically poured into a glass. This action destabilizes the dissolved gas and acts as a catalyst for nucleation of the carbon dioxide bubbles. As the bubbles rapidly form they rise to the surface of the beer due to a difference in density with the liquid. The bubbles will accumulate as a layer of foam on top of the liquid if the rate at which they are forming exceeds the rate at which they are breaking. Once the beer has finished being poured, the rate of formation slows considerably and eventually the foam layer will be perceived as relatively stable, although ultimately disappearing either because the beer has been consumed or all the bubbles have broken.

The key factor as to the stability of the bubbles are the characteristics of the liquid film that surrounds the entrapped gas, specifically the presence of various surface-active components of the beer. As illustrated in Figure 8.8, compounds that are referred to as "surface-active" tend to accumulate at gas-liquid interfaces because they have a chemical structure containing both hydrophilic (water loving) and hydrophobic (water hating) parts.

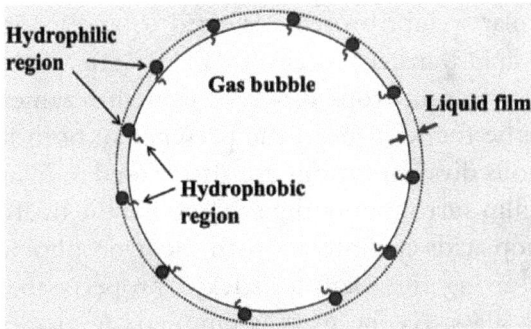

Figure 8.8. An illustration of a bubble of carbon dioxide in an aqueous environment with surfactant molecules accumulating in the liquid film (thickness is not to scale). The hydrophobic portion of the molecule will favor the gas phase.

Therefore, being at the gas-liquid interface satisfies both tendencies and is thermodynamically favorable. Surfactants affect the liquid film surrounding the bubbles by lowering the surface tension, thereby improving foam stability. Bubble size is also important as larger bubbles are much less stable than small bubbles. The foam is initially composed of mostly small bubbles, however, some bubbles tend to grow as gas diffuses from smaller into larger bubbles. This diffusion and associated bubble growth occur much slower with nitrogen than with carbon dioxide because of nitrogen's lower solubility. Thus, when nitrogen is used as a portion of the "beer gas," for example with Guinness draft, the foam is composed of smaller more stable bubbles with higher foam stability.

The soluble proteins generated in the brewer's wort during mashing are soluble in part due to their conformation. The hydrophilic amino acids are exposed on the protein surface, while the hydrophobic region is isolated on the interior. During boiling these proteins tend to denature, losing their native conformation and exposing the hydrophobic regions. Solubility is reduced and some will end up in the trub, but a fraction will remain in the beer and exhibit their surfactant properties by helping to stabilize the foam. The most important foam stabilizing proteins originating from the barley seem to be a high molecular weight protein, referred to as Protein Z, and the much smaller lipid transfer protein, LPT1.

In addition to numerous different protein fragments and peptides that can be found in beer, the presence of both hop bittering acids and various divalent cations are also related to foam properties. In the liquid film surrounding the gas bubble, the hydrophobic portions of the hop acids can interact with the hydrophobic portions of the proteins, giving the foam a "sticky" property that helps it to adhere to the glass. As the glass is emptied, an attractive lace-like appearance can be observed on the inside wall (see Figure 8.9). This hydrophobic interaction is strengthened in the presence of certain divalent cations such as zinc, which presumably reduces the effects of the negative charges on the iso-α-acids. With all the inherent variability in water composition, recipes and brewhouse operations, it is likely that the characteristics of the foam are at least somewhat unique for every beer.

It is possible to quantify foam stability for any given beer, either through the use of a simple laboratory apparatus and manual measurements or with the use of automated commercial instruments, for example as provided by Haffmans or Steinfurth. Although there is no single standardized test for foam stability, the principles are much the same. Foam is generated in the beer sample, typically by pouring or

Figure 8.9.   An example of lacing on the inside of a glass of lager.

agitating, and then the rate and/or quantity of foam collapse is measured. In the ASBC Beer-22 Method A, referred to as the Sigma Method, a separatory funnel is used to capture the volume of foam collapsed over a 200 second time period. A calculation is made of the foam stability index, or sigma value, based on the relative volume of foam collapsed compared to the foam remaining. Foam stability is only a relative concept, but if the same method for testing is used on all samples, then changes based on different recipes, processing or storage conditions can be assessed.

8.8 Topics for Discussion

- Approximately what accuracy of % ABV could you expect to achieve if it is calculated based on change in specific gravity?
- How consistent would you expect the wort composition to be if the source of malt changed? What is most likely to be different?
- Is the taste and aroma of beer largely a subjective experience or can a trained panel be purely objective in its assessment?
- What off-flavors in beer are easiest for you to detect and how common are they in craft beer?
- How important is it to store beer under refrigerated conditions and what are the consequences of exposure to high temperature, for example during pasteurization?
- Considering that light American lager is the largest selling beer in the world, how important do you think foam stability is as a quality parameter?

8.9 Further Reading

1. American society of brewing chemists methods of analysis. 2020. www.methods.asbcnet.org/toc.aspx#Beer
2. Bamforth CW, Foam ASBC. *Handbook Series Practical Guides for Beer Quality*. St. Paul MN, USA: American Society of Brewing Chemists; 2012.
3. Boulton C. *Encyclopaedia of Brewing*. Wiley Blackwell; 2013.
4. Langstaff S. The defects wheel for beer. 2009. www.defectswheel.com

https://doi.org/10.1142/9789811225321_0009

Chapter 9

The Craft Beer Industry

9.1 Introduction

In the late 1970s and early 1980s, it appeared that the United States beer market was on its way to being dominated by a few large national players. The number of brewing companies had dipped below 50 and the number of breweries below 100. The business and economics literature of the time focused on economies of scale and minimum efficient scale, arguing that brewing below millions of barrels no longer made sense in a rapidly consolidating marketplace. It was at that point that a few radical pioneers started to think and brew differently, starting a shift in the marketplace that has ultimately seen small breweries number in the thousands (~7,500 breweries in the United States and growing, as of mid-2019) and take double-digit share of the market by volume, even facing some of the largest and most efficient brewing companies in the world.

This chapter details the rise of those small craft breweries, their strengths and success, as well as some of the challenges they face in entering the market, becoming profitable, and turning their passion for brewing into sustainable businesses.

As with any market populated by thousands of players, it is not a singular story, and so much of the story will be about the dispersion of models as much as the central tendency. Similarly, the tremendous variation over time and place requires rooting the craft industry in its historical, regulatory and broader economic context. It is only by understanding the unique conditions that bred and shaped craft brewers that their current success and challenges fully take shape.

9.2 Craft Begins

There are many dates that could be linked with the beginning of small brewing in the United States, but two typically stand above the others in importance. The first of these events was Fritz Maytag's purchase of the Anchor Brewing Company in 1965, and his subsequent decision in 1969 to reformulate Anchor's offerings to all-malt beers. Throughout the 1970s, Maytag continued to drive Anchor Brewing to revamp its offerings, mixing European traditions with distinctly American updates and ingredients. A primary example is Liberty Ale, first brewed in 1975, and often viewed as the first modern American IPA, now the most popular style from American small brewers.

The industry as a whole gained a boost in 1976, when lobbying by the Brewers Association of America helped achieve a reduction in excise taxes for small brewers, lowering the Federal excise tax rate on the first 60,000 barrels of production for brewers producing less than 2 million barrels (1 barrel = 31 gallons). All brewers pay additional taxes, known as excise taxes at both the Federal and State level (and some even at the local level), so a reduction in these taxes for small producers is vital in helping offset some of the challenges in competing against scale competitors with significantly lower cost of production per unit.

It was also in 1976 that Jack McAullife founded New Albion Brewing Company in Sonoma, CA. New Albion is commonly referenced as the first modern microbrewery. McAullife, an engineer and homebrewer (which was still illegal), built his equipment from scratch or out of old dairy equipment as there was no market for small brewer equipment or American suppliers.

The potential for more microbreweries was increased in 1978 when President Jimmy Carter signed H.R. 1337 (containing an amendment by Senator Alan Cranston of California), legalizing homebrewing at a Federal level in the United States. Although home wine making was legalized during the repeal of Prohibition in 1933, home beer making had continued to be banned until H.R. 1337 went into effect on February 1, 1979.[1]

This allowed the rapid development of a homebrewing supply industry in the United States. In turn, the homebrewing industry created ties with and propelled the opening of new small breweries. For instance, in 1979, Ken Grossman and Paul Camusi used equipment and knowledge from their homebrew supply shop to found Sierra Nevada Brewing Company (now the 7th largest brewing company in the United States).

The nascent small brewing industry was given yet another boost in the early 1980s, when states began amending tied-house laws and allowing on premise sales for small breweries and brewpubs. Tied-house laws, designed to prevent some of the abuses that had led to prohibition, prohibited connections between production and retail, effectively outlawing the direct sale of beer by brewers, as well as many business models such as the brewpub. Washington led the way in 1982 and California almost immediately followed. The increased market access of direct sale of beer to customers was another key piece in getting many small brewers off the ground, allowing them to compete in new channels that didn't require the scale of the large brewers.

Demand

Who were the consumers of those early craft beer pioneers, and what are the demand markers that drive the craft market today? Reading the stories of the early craft pioneers, and compiling data from consumer surveys on craft purchases today, common themes emerge. Craft has long been built on demand for flavor, variety and values.

When asked what characteristics are important to them when choosing a craft brewed beer to purchase, flavor is typically the number one

---

[1] The timeline of state legalization is more difficult to define. Although many states had explicitly banned homebrewing and so have definitive bills to point to for legality, others simply treated the federal legalization as a de facto state legalization. Consequently, there is not a specific state bill to point to for legalization in many states. If fact, in some states, there are some that argue that homebrewing has not been legalized, although it is now generally considered a legal activity (with varying rights) in all 50 states.

answer. In a recent Nielsen survey of craft purchasers, 99% cited flavor as a reason they purchase craft. Whereas in the 1980s, the vast majority of the U.S. beer market was dominated by American lagers and light lagers — beers typically built around very minimal flavor and with refreshment as the primary value proposition, craft, along with specialty imports, reintroduced Americans to flavor as a primary reason to drink beer.

Secondly, many craft consumers seek variety. The early craft drinkers were often looking to recreate traditional beer styles they had tried while traveling abroad, to European countries like Germany or the United Kingdom; beers which weren't available in the U.S. marketplace. Today, variety has taken on a broader meaning. Craft serves a wide variety of occasions, and craft often offers specific flavors for those occasions. In addition, many craft consumers seek variety for variety's sake, a phenomenon that has been amplified by technology tools such as rating sites.

Finally, craft has long incorporated a value component into its demand mechanism. Whether "local" or "independent" or simply "small business," craft coincided with a shift in American food culture looking to buy and shop local as well as understand where food and beverage products came from. This isn't unique to beer, and in many ways craft brewing has paralleled the rise of specialty, artisanal and local products in a variety of other industries. Think farmer's markets, local coffee roasters and the rise of an organic food industry.

Premiumization

Craft is not the only part of the beer business that has moved upmarket as beer drinkers seek more from their beer. The beer business as a whole is currently experiencing what is often referred to as a "premiumization" trend, or a period where beer lovers trade up across the beer aisle as shown by the data in Table 9.1. This isn't the first premiumization trend beer has seen, as the dominant brands of today rose to prominence in a similar trend — hence the reason those brands are often referred to as the "premium" category.

Figure 9.1 shows these trends over time, using a combination of data from R.S. Weinberg, Beer Marketer's Insights, and the Brewers Association to show the volume share of three segments of the beer market from 1950 to today. As you can see, the rise of the upper end of the beer market going

Table 9.1. The premiumization of the beer market between 2011 and 2016.

Brands (share of $ sales)	2011	2016 (YTD)	Change
Brands priced below leading four	24.0%	18.5%	–5.5%
Leading four brands	44.6%	37.0%	–7.6%
Brands priced above leading four	31.4%	44.5%	13.1%

Source: IRI Group, MULO+C YTD (through 11-06-16)

Beer's Premiumization Trends

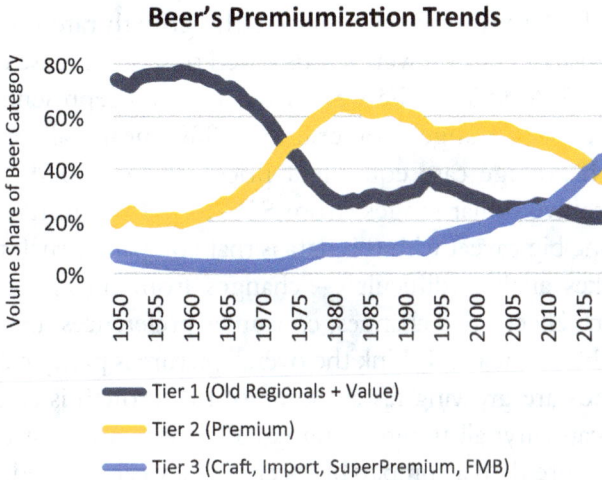

Tier 1 (Old Regionals + Value)
Tier 2 (Premium)
Tier 3 (Craft, Import, SuperPremium, FMB)

Figure 9.1. Trends in the premiumization of the beer market from 1950 to the present.

on today isn't that different from the trend from 1960–1980, where the current dominant brands took over from previous market leaders.

These trends would be even more accentuated if they were presented in value terms, where the higher price point brand share is growing even faster than in volume terms. For example, the table above presents scan data — sales data captured using bar codes as supermarkets — convenience stores and more — for every beer brand for the full year of 2011 and 2016 YTD and the changes are remarkable. Although the dominant four brands still account for the plurality of volume and dollar sales, their contributions have rapidly declined as the market has shifted upward under their feet.

The Table 9.1 above shows the share of dollar sales in scan in those two years. In 2011, the largest four brands were 44.6% of dollar sales and all brands priced above them were 31.4%. By 2016, the leading four brands had declined to 37.0% of dollar sales whereas brands priced above them were now 7.5 share points larger, at 44.5% of total dollar sales YTD.

Premiumization within Craft Beer

Premiumization doesn't just occur across beer categories, but within them as well. A third price chart shows the growth rate for just small and independent craft brewer brands based on average case equivalent prices from $26–$65 — the size of the bubbles represents the case sales size for each range. For example, $30 means all of the craft brands with average case equivalent prices of $30–30.99. $50, $55 and $60 are five-dollar ranges (from $50–54.99, for instance).

Now one big caveat with this data is that comparing such tight price ranges makes analysis difficult — changes from one price point to another can reflect price changes, consumer preferences, brand quality and more. Nevertheless, I think the overall picture is pretty clear: higher priced brands are growing faster. Some of this growth is certainly due to reverse causality: all things being equal, a beer with a strong brand can charge more than a comparable beer with a weaker brand, and separating the direction of those causal arrows isn't simple. Some growth is also likely due to smaller bases as the price moves up, but the percentage growth trends are pretty clear, and the volumes from $35–$45 have become a significant portion of craft's overall volume.

As you can see on the graph (as shown in Figure 9.2), much of the struggle for craft as a category this year has occurred in the $25–$35 range, which is an increasingly crowded segment of the beer market. In addition to a growing number of craft brewer brands, that range includes the growing premium plus brands from the large brewers, newly acquired brewers that are being scaled up, many import brands and more. Brands that can position themselves above that range have found more greenfield space and have been on average more successful (albeit at lower volumes), as seen by the aggregated figures.

What does this mean for craft brewers? Both the recent data and the longer run historical experience show that a brand's premium

Craft Prices and Growth

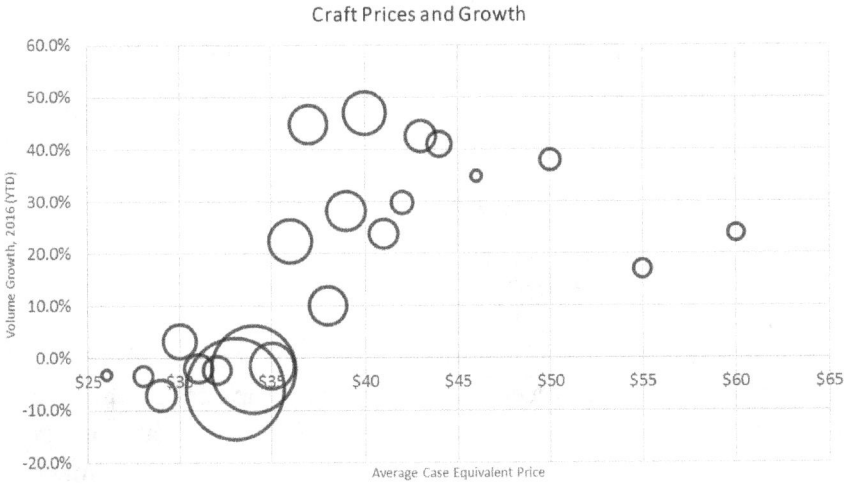

Figure 9.2.   Volume growth in craft beer based on case (of 24 × 12 oz.) equivalent price.

Source: BA Craft, IRI Group, MULO+C YTD 2016 (through 11-06-16)

positioning is constantly under pressure, and brewers must constantly work to add value to their brands through quality, consistency, flavor, marketing and customer interaction. The premiumization wheel has already turned multiple times in the beer industry and will continue to turn in the future. What the next cycle will bring is anyone's guess, but brewers who proactively recognize that challenge and focus on what they can control (quality and consistency, for instance) will be better positioned to ride the wave rather than watch it crash over them.

Demographics

As part of the above premium segment of the beer market, it's no surprise that craft's demographics are different than that of a typical beer drinker. So what are the demographics of the craft market and how do they differ from the overall population and overall beer market?

Let's start really broadly. If we use drinking "at least several times a year" as our standard, around 40% of the 21+ population is now a craft drinker (source: Nielsen-Harris on Demand). That's been going up about one to two percentage points per year as shown in Table 9.2.

Table 9.2. The percentage of the population that can be considered as craft beer drinkers.

2015	2016	2017	2018
35%	37%	38%	40%

Source: Percentages are based on several waves of data from surveys fielded online by The Harris Poll between 2015 and 2018

The United States 21+ population has been going up by ~2.5M legal drinking age adults in recent years, so that means craft is getting a slightly bigger bump than that every year (since the category gets ~40% of those new 2.5M + 1–2% more of the total ~240M drinkers). That's likely averaged around 4–5 million new craft drinkers a year using a "several times a year" drinker definition.

If we look at more frequent craft consumption, not surprisingly, the numbers drop. Scarborough (another division of Nielsen) estimated that in 2017, 7.3% of 21+ adults had been a craft drinker in the last month. That's about 17.5M people in craft's core.

Socioeconomic Status

Beverage alcohol consumption tends to draw from a demographic that is slightly wealthier than the overall population. This may not be surprising. Compared to expenses like food or rent, beer, wine and spirits are nonessential, and so Americans tend on average to consume more of them only when those basic economic needs have been met.

For craft brewers, their higher priced point products draw even more heavily from the upper end of the socioeconomic spectrum. The table below (Table 9.3) (from Scarborough Research) shows indices of craft purchasing by household income. An index of 100 means that group purchases the same amount as their percentage in the population. Not only is craft skewed to higher incomes, the index increases with income. So for example, even though households that make $100,000 or more only represent 22.5% of the U.S. adult population, they make up 42.5% of craft purchasers.

Table 9.3. The purchasing index for craft beer drinkers based on household income.

Household income	Index
Less than $50,000	47.1
$50-99,999	114.8
$100,000+	189.7

Source: Scarborough Research

Gender

Next, let's look at gender. Taking the broad, "at least several times a year" view, craft drinkers are 31.5% female and 68.5% male in 2018 (source: Nielsen-Harris on Demand). That's pretty much the same as monthly, where Scarborough found 31.1% female and 68.9% male. Over time, as craft has grown in the marketplace, these numbers are moving closer to 50-50.

In 2015, the same Harris Poll found "several times a year" craft drinkers were 29.1% female and 70.9% male. That's 2% points in shift toward females, in a three year period where total craft went up ~5% points in the country. When you add that all up, it suggests that from 2015 to 2018, craft has added ~14.7 million drinkers, of which a bit below half (~6.6M) were women. If that data is correct, craft is now onboarding men and women into the category at roughly their percentages in the population.

Want to check that math?

In 2015, there were 234,380,464 21+ adults (Census Bureau) and 35% drank craft beer (Nielsen- Harris on Demand). That's .35*234,380,464 = 82,033,162 craft drinkers. In 2018, we're estimating that at 241,876,792 21+ and 40% (Harris), so 96,750,717, or 14,717,554 more than in 2015. For female craft drinkers, it's gone from .291 * 82.0 M = 23,871,650 to .315 * 96.8M = 30,476,476, or +6.6M (45% of the total) from 2015.

These demographics can vary a lot based on local market conditions. Looking at individual markets, Portland Oregon's craft drinker breakdown is 52.7% female and 47.3% male (source: Scarborough), so there will be a huge range depending on the condition of a particular market and its own unique demographics.

9.3 Geography

As the number of breweries in the United States has exploded, from just above 1,500 in 2008 to almost 7,500 a decade later, the number of cities and towns with a brewery has also increased. Today, roughly 85% of Americans live within 10 miles of a brewery, a return to a previous era where the vast majority of Americans have ready access to fresh, locally made beer.

This ubiquity of breweries might make one assume that breweries are fairly evenly distributed around the country. However, this is not the case, and there are interesting variations in the location of breweries by place, state and city size that provide insight into how and why breweries choose to locate.

Breweries by State

Let's start at the stating something obvious. The number of breweries generally follows population. The state with the most breweries in the U.S. is California, and behemoth states like New York, Texas and Florida are all in the top 15 states in terms of absolute numbers of breweries. As we move to looking at breweries per capita, however, more variation emerges, and the story becomes less clear. For example, Vermont, a small and not particularly dense state, has approximately five times as many breweries as New York controlling for population. Some of this is related to things like costs of operation, but more broadly, we see that particular states such as Colorado, Oregon, Vermont or Maine, have much denser concentration of breweries than other states. In fact, U.S. states have more variation in breweries per capita than do European countries!

So, what explains this tremendous variation, exceeding that of Europe, with its much greater cultural and economic differences? To deepen the puzzle, the differences in producer location have largely persisted over time, despite a national consumer market that shows many signs of common development and regulations that have seen many signs of convergence. Looking back nearly two decades, from 2000 to 2017, the number of brewery licenses (a good measure of breweries plus breweries in planning) grew 4.4 times from 1,964 to 8,657. However, the relationship between licenses per capita in 2000 and 2017 is nearly perfectly linear (r-squared = 0.923). So in an era where a minimum of 6,693 firms made location decisions, why did they locate in largely the same patterns as the first 2,000?

At the state level, the answer boils down to a combination of consumer preferences, legislation/regulation, and the interaction between the two. Because of the 21st Amendment to the constitution, the regulation of beverage alcohol is largely left to the states, creating tremendous variation in how breweries can operate in the marketplace. We'll cover some of these rules in more depth when we look at brewery business models later in the chapter.

These laws and rules meant that it was, and still is, much easier to run and operate a brewery in some states than others. That created early mover advantages for breweries, allowing beer drinkers in those states to develop a taste for fuller-flavored beer. In addition, this often created a self-perpetuating cycle. States that had breweries saw those breweries grow and look to expand their market. They lobbied for even more market access and even better rules. In doing so, they created market opportunities for further entrants, which in turn increased the economic and political power of the brewing sector, continuing this cycle over time.

Within States

So, if state-by-state variations are explained by laws and regulations, why do some cities or regions within states have greater brewery

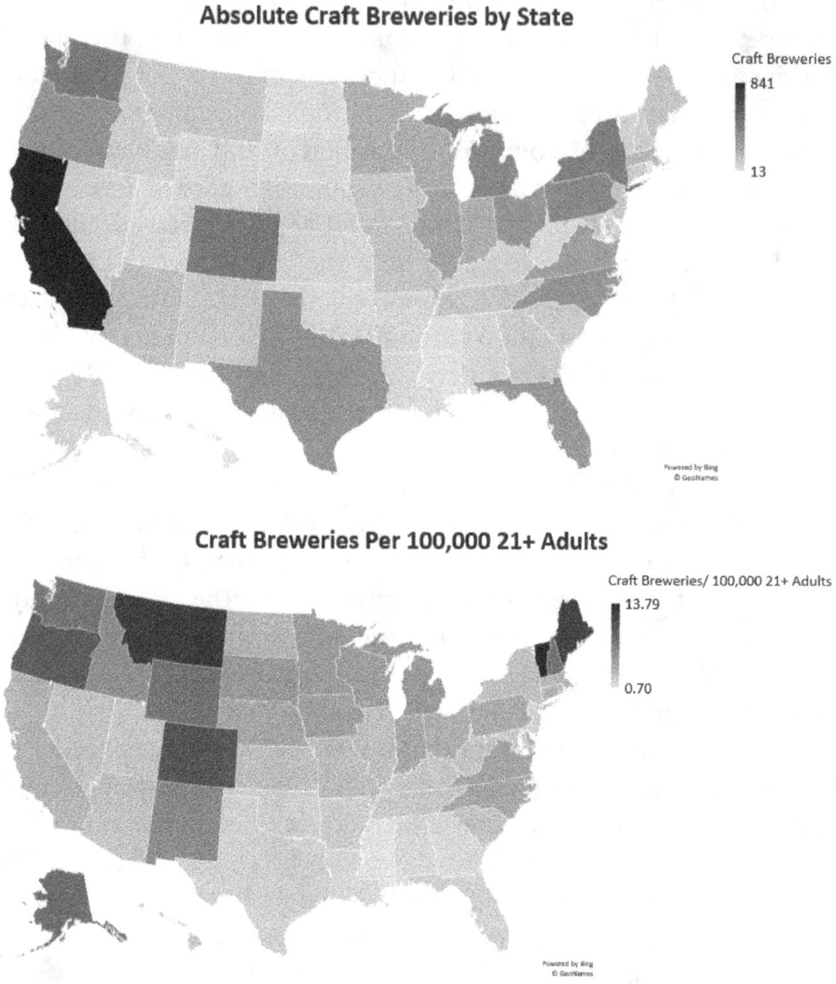

Absolute Craft Breweries by State

Craft Breweries Per 100,000 21+ Adults

Figure 9.3.   Number of craft breweries by US state (top) and per 100,000 inhabitants (bottom).

Table 9.4. The number of breweries per capita based on the size of the urban area population.

Pop (2010 Census), Urban areas	2013	2018	% Ch
<5M	0.45	1.26	183%
1M–5M	1.06	2.51	136%
100K–1M	1.11	2.77	149%
10K–100K	1.70	3.87	127%
2.5K–10K	2.11	5.00	137%
Not in a 2010 urban area	0.44	1.00	129%
Non-urban as % of Total	8.9%	8.4%	

Sources. US Census Bureau, Brewers Association

density than others? Some of this is demographics. As discussed earlier, cities with higher income and younger populations are going to have a market that is more conducive to small breweries than those with older or poorer populations. But the size of cities also appears to matter.

The table (Table 9.4) above compares breweries per capita by the size of urban areas around the country.

This uses Census Bureau data on all urban areas above 2,500 people. The results are fascinating, and show that in per capita terms, you actually get more brewery density in smaller towns. Why might that be?

Much of this can likely be explained by costs (land and labor) and competition. Not only are larger cities typically more expensive, they are already going to have a very competitive landscape of food and beverage options. In contrast, smaller cities may have provided more greenfield sites for small breweries to thrive.

These patterns have remained largely steady over time, though the number of breweries in rural areas and the smallest towns under 2,500 people has actually decreased as a percentage of all breweries since 2013, despite a growth of 129% in the number those breweries overall (note that total brewery growth has been 142% since 2013).

Within Cities

Finally, there is now a growing literature looking at where breweries locate within cities. Much of the recent scholarship on urban geography and brewery locations has focused on evidence that breweries cluster (see[2,4] for broader context on economic clusters). Similarly, looking at brewery location by census tract,[6] write: "the strongest predictor of whether a craft brewery opened in 2013 or later in a neighborhood was the presence of a prior brewery."

There are a number of reasons why breweries might choose to colocate. The first is again demographics. Breweries have a specific clientele they are looking for that might be found in specific parts of cities. Another is risk-aversion. Choosing a location where a successful brewery already operates reduces risk in choosing a poor location — even if it may increase competition. Combining these two arguments, there may also be advantages to colocations. Consumers may seek out brewery districts over individual breweries, or be more likely to extend a first brewery visit to another location if the second brewery is nearby. A final set of arguments is related to city planning and zoning. As a business that involves both beverage alcohol and manufacturing, breweries may have limited choice when choosing acceptable business locations. The answer is likely a combination of all of these reasons driving breweries to colocate within cities.

9.4 Business Models

Craft brewers use a variety a business and distribution models. Broadly, those choices fall along a continuum from a distribution focused model — selling beer through on- and off-premise retailers, often through independent distributors — to a onsite model — selling beer directly at the brewery via a brewpub or taproom. Most small breweries operate a mixed model, with some onsite presence as well as beer sold through distribution outside of the brewery.

Distribution

Within the distribution model there are in-turn numerous business model choices that brewers face, including how to distribute beer as well as what geographic region to distribute in.

Once a brewer chooses to distribute their beer, they will need to choose how to get that beer to market. The vast majority of beer in the United States travels through an independent distributor who purchases beer from a variety of breweries and then sells that beer to retailers for final consumer sale. Many of those distributors are aligned with one of the large brewing companies, allowing them the scale to access the vast majority of retailers. Signing with the right distributor is an important choice for any brewer, as many states have what are known as "beer franchise laws" that lock a brewer together with their distributor once they have signed a distribution agreement.

As states have liberalized their distribution laws, many craft brewers have also begun self-distributing their beer, i.e., acting as their own wholesaler. Although this isn't legal in all fifty states (self-distribution is currently legal in 39 states), self-distribution often provides brewers an avenue to demonstrate market demand for their product before signing with an independent distributor, as well as gain greater understanding of how distribution works. Some brewers choose to keep a portion of their distribution footprint under the brewery's control even as they expand — often the distribution territory surrounding the brewery. In rare cases — and where legal — craft breweries will expand their distribution arm into a separate distribution business and begin distributing brands from other small breweries in addition to their own brands.

Channels

Historically, the beer industry has conceptualized two major channels for selling beer to consumers: on-premise (so named because you drink beer on the premises) and off-premise (where you buy beer to take away and drink off the premises). On-premise is bars and restaurants, whereas off-premise are retail stores like grocery, convenience

Table 9.5.   Relative draught versus packaged beer production for different sizes of breweries.

Category	Size (bbls)	Production breakdown	
		Draught	**Packaged**
Production Breweries	1–1,000	86.4%	13.6%
	1,000–5,000	59.8%	40.2%
	5,000–15,000	53.5%	46.5%
	15,000–60,000	41.6%	58.4%
	60,000+	37.3%	62.7%
Brewpubs	1–1,000	86.0%	14.0%
	1,000+	63.6%	36.4%

Source: Brewers Association Brewery Operations Benchmarking (2016)

or big box stores. Volume-wise the overall beer market is driven much more heavily by the latter — with around 80% of the total market coming from off-premise by volume.

Craft brewers, however, have typically focused as much, if not more on on-premise channels. There are a multitude of reasons for this, but many boil down to scale economics. Off-premise retail channels tend to be larger and consequently look for larger supplier partners when building the list of brands they plan to sell. In addition, many brewers simply choose not to invest in the manufacturing infrastructure required to put beer into packages designed for off-premise retail stores. The table below (Table 9.5) shows how this mix shifts on average as breweries grow.

Onsite

Because most breweries are quite small — ~75% of the breweries in the country make 1,000 barrels or less — the vast majority of brewery business models incorporate some direct to consumer sales at the breweries. Broadly we can separate onsite brewery business models into two categories based on their service model: brewpubs, which also operate a restaurant, and taprooms or tasting rooms, which simply serve beer. In reality there are numerous variations on these patterns, often influenced by state laws that govern what breweries can sell at their brewery and how they can sell it. Examples of these rules include selling beer from other breweries, selling wine and spirits, selling beer to go and limits on sales of beer.

At the beginning of the craft beer revolution, most onsite breweries were brewpubs. This was by necessity. Selling beer at your brewery typically required holding a brewpub license. As the laws governing direct to consumer sales have liberalized, however, more breweries have shifted into the taproom model, focusing their efforts around selling beer without the added complications and costs that come from adding a kitchen and staffing for food service.

Using government data, we can see how much these direct at the brewery sales have grown over time. The Figure 9.4 shows data from the Tax and Trade Bureau (T.T.B.) the government body tasked with regulating breweries as well as collecting excise tax revenue. It shows "premise use" production, which measures sales directly at breweries from tax determined tanks or portions of the brewery, and is a good proxy for total at the brewery sales. As you can see, these sales have exploded in recent years, from just over 500,000 barrels in 2013 (about 1 out of 400 beers sold in the US), to more than 3 million barrels in 2018 (about 1 out of 65 beers sold in the U.S.).

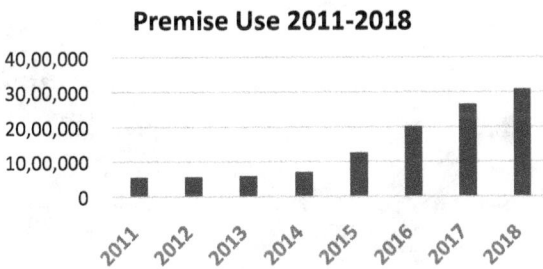

Premise Use 2011-2018

Figure 9.4. The trend in the number of barrels of beer sold on brewery premises from 2011.

Source: Tax and Trade Bureau

9.5 A Competitive Market

As with any industry, the success of the market has meant new entrants and rising competitiveness. Competition has come from all directions, from the world's largest brewers to small local players. In many ways, these challenges suggest the future of craft, as it becomes both more global, and more local.

The Large Brewers Respond to Craft

As we think about both the current state of the market, and its future, it may be helpful to think about the development of the craft brewing industry within the larger context of industry growth, consolidation, and fragmentation. A study published in the Harvard Business Review in 2002 highlighted the four stages of industry consolidation and fragmentation. At the time, they placed brewing at the end of the scale stage, moving into the focus stage (though the data for this chart is from much earlier than 2002).

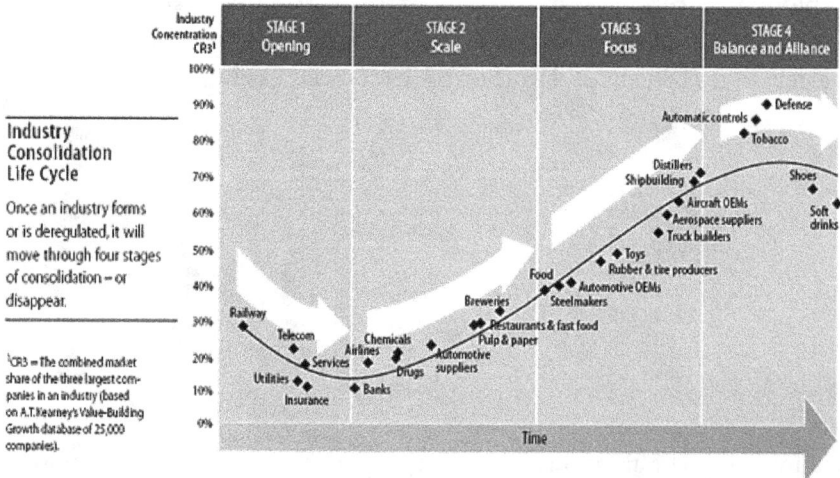

Figure 9.5.   Industry consolidation depending on the stage of the business life cycle, putting breweries at Stage 2 at the end of the 20th century.

Source: https://hbr.org/2002/12/the-consolidation-curve

Over the next decades, breweries continued to consolidate and focus on the center of the market: American lagers and light lagers. This culminated in 2008 with the merger of Miller and Coors, the same year Anheuser-Busch was acquired by the global brewing behemoth InBev. This process of consolidation made the large brewers incredibly efficient and effective at selling a particular type of beer, and in the process, they built enormous brands. At the same time, it created opportunities at the margins of the industry, niches where small-scale competitors could focus on small-scale opportunities and consumers that the large brewers were ignoring as they consolidated and scaled.

This is where the craft industry had its roots and found success. Small, ignored styles of fuller-flavored beer. Retail channels that didn't have much scale, such as specialty retailers, high-end on-premise establishments, and more. Over time, however, craft brewers started to achieve real scale for themselves as they built on these channels and opportunities. That brings us to the fourth stage: balance and alliance. The market share of large firms can dip as the collective efforts of small, nimble competitors eat at the edges of their market.

They respond by forming alliances, acquiring those small startups, and trying to protect their flanks while still focusing on their core business. This has occurred in fits and starts over the years. When craft saw tremendous growth in the mid-1990s, the large U.S. brewers both acquired some firms (such as Miller acquiring stakes in Shipyard and Celis, both of which they later divested in) or building their own in-house brands to compete in those markets. It has reached another level in the past 5 years, with numerous acquisitions by every major player in the U.S. beer market. The table below summarize some of the acquisitions and in-house brands built by the large brewing companies in the United States. It is likely that these companies will have purchased more brands, and perhaps even divested some of these brands, by the time this book is published.

Anheuser-Busch:
Brands: Shock Top, Landshark, Ziegenbock
Acquisitions: 10 Barrel, Blue Point, Breckenridge, Devils Backbone, Elysian, Four Peaks, Golden Road, Goose Island, Karbach, Wicked Weed, and Wild Series brands

Partially ownership of Craft Brew Alliance, including Kona, Omission, Red Hook, and Widmer Brothers brands

MillerCoors

Brands: A.C. Golden, Batch 19, Blue Moon, Colorado Native

Acquisitions: Hop Valley, Revolver, Saint Archer, and Terrapin

Constellation Brands

Acquisitions: Ballast Point, Four Corners, Funky Buddha

Heineken USA

Acquisitions: Lagunitas

Sapporo

Acquisitions: Anchor Brewing

FIFCO USA

Acquisitions: Genesee, Mactarnahan's, Magic Hat, Portland and Pyramid brands

Craft Brewer Responses

Similar acquisitions have happened in nearly every developed brewing market, including the United Kingdom, Canada, Belgium, Ireland,

Belgium

UK

Canada

USA

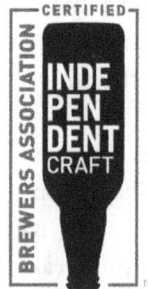

Australia and more. In response, numerous small brewer trade associations have begun launching consumer awareness campaigns and creating certified marks for distinguishing small brewer brands from acquired brands being scaled up by large global brewers. Some examples include:

Competition From Local

Competition for the craft brewing market hasn't come solely from the large brewing companies, but also from within. With the proliferation of breweries, a market that used to be characterized by a rising tide raising all breweries has become one where craft brewers must find their niche to succeed. A primary challenge is how to grow in an era when many craft consumer prioritize local in their purchasing decisions. As seen in the chart below, local is more important in the craft market than in other parts of beverage alcohol.

As the demand growth of craft has slowed, this has meant that much of the remaining growth has funneled into local brands at the expense of national craft brands. Finding the balance between economies of scale and a distribution footprint that stays connected to demand is going to be a central challenge for many regional craft brewers in the coming years.

9.6 The Future of Small Brewing in the U.S.

As craft brewing hits 50 years since Anchor Brewing reformulated to all malt beers, the gap between the modern industry and those humble beginnings has become a chasm that grows daily. Thousands of breweries with a diversity of business models now compete across every major urban area. This creates a contradiction. The segment of fuller-flavored local beer is larger than ever before, with more drinkers and higher sales. At the same time, the market is more competitive than ever before, making it more difficult for any individual company to realize the gains from this growing market.

The very success of the industry has generated a market where it is more difficult for any brewery to stay relevant in a world; one where the definition of local continually tightens and the number of

How important is "local" in purchase decisions?

Nielsen Surveys: Sum of very/somewhat important

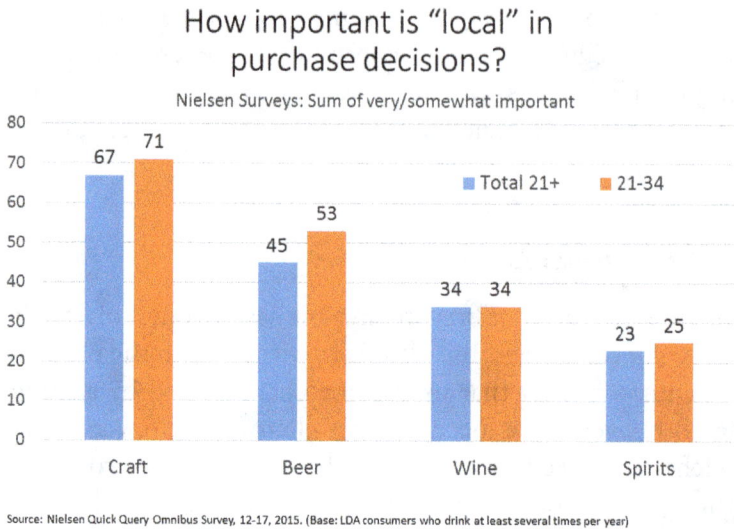

Source: Nielsen Quick Query Omnibus Survey, 12-17, 2015. (Base: LDA consumers who drink at least several times per year)

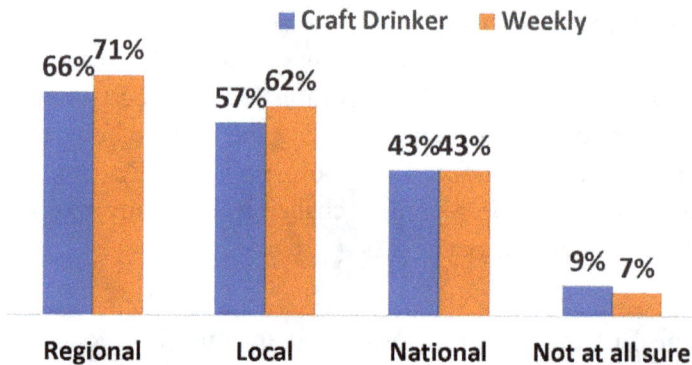

Figure 9.6.   The importance of being local in the alcoholic beverage industry, especially for craft beer drinkers (top) and the popularity of regional and local brands versus national brands (bottom) especially for regular (weekly) drinkers.

Source: Brewers Association Craft Beer Insights Poll (CIP) conducted online by The Harris Poll/Nielsen in May 2019 (n = 1,100 Craft Drinkers, n = 480 Weekly Craft Drinkers Figure

breweries needs to be updated weekly. Given this contradiction, what are some strategies for individual breweries to succeed in this modern era? Although any business strategy needs to be optimized for an individual company's location, competencies, and identity, there are common lessons that mirror those seen in other competitive consumer product categories.

The first is a concept that is simple in theory and difficult in execution: differentiation. Although a set of regional craft brewers have likely reached the critical scale necessary to survive in broad distribution, allowing them to compete on efficiency, speed to market, price and lower margins, most small brewers will continue to rely on their ability to charge a premium to offset their smaller scale and higher costs of production. Some of this ability is tied to continued consumer demand for small and independent local businesses. However, as intimated in the opening of that section, small and local are only partial differentiators. When there are a plethora of small and local options, why should a beer drinker choose the beers from a particular small local brewery?

Consequently, the success of most small breweries in the future will be directly tied to their ability to create a unique identity in their local beer market. The paths to doing so are varied; differentiation can take the form of service model, beer quality, beer styles, marketing and branding, target audience, location and more. The breweries that are able to do this on the most dimensions will stand out and thrive. Those that don't, will increasingly struggle to compete in a world where they are offering commodities in a world demanding unique products.

An offshoot of this strategy are breweries attempting to differentiate via their raw material inputs: incorporating local ingredients or even taking on the farming and processing of the agricultural products in their beer. Some of this has been encouraged by U.S. States, such as New York, which created a license that provided brewers additional retail rights in exchange for using a higher percentage of local ingredients. Others are doing it simply as part of their business model and brand, either incorporating all local ingredients or growing their own inputs.

A second strategy that many breweries have already begun exploring is product diversity. In many ways, diverse product and service offerings was an idea baked into many of the earliest small brewers in the form of the brewpub. Today's breweries are taking this idea to the next level. One dimension involves product diversity, building on a breweries brand to expand into tangential alcoholic beverages, nonalcoholic beverages, or even other consumer goods products. In states where legal, breweries are producing wine, spirits, hard kombucha, cider, mead, hard seltzers and more. Many of those products require similar production equipment, allowing breweries to ensure that their capital investments continue to provide revenue streams. Nonalcoholic beverages and products are example of what is old is new again, as many breweries used similar product extensions to survive prohibition. Some small breweries are taking this even further, and expanding not just their product, but also their service offerings. Craft beer hotels or non-brewing retail locations offer an opportunity for breweries to expand their service brand.

To close, the challenges of the craft brewing market today do not represent a bubble bursting or a fundamental collapse in demand. The U.S. beer market, like many other around the world, has evolved, and consumer demand for more flavor, variety and local options does not appear to be receding anytime in the future. At the same time, a rush of entrants has made this space increasingly competitive. The future of the industry, therefore, involves many of the same strategies that kicked off craft brewing in the first place: standing out from the crowd and finding ways to be different. Unlike the early days, however, these strategies must now be much more fractured, tailored to particular firms and nuanced. The successful firms of the future will need to think like the early pioneers, and try new things not currently in the marketplace. If they succeed, the future of U.S. beer will continue to be bright.

9.7 Topics for Discussion

- What other consumer product markets can you think of with similar demand trends to beer? Why do you think there are comparisons? Where is that market in its development relative to beer?
- What are the limits of premiumization in beer? Why might they be higher or lower than in another market (e.g. wine)?
- What do the demographics of craft suggest about future growth? How would you expect those demographics to change going forward? Why?
- What unique regulations for beer exist in your city/state? How do you think those have shaped the development of your local market?

- Which rules do you think are most important in shaping the growth of the beer sector? Why?
- Which would you change from a public policy perspective? Why? What would be the benefits and costs of such a change?
- What opportunities and challenges does the growth in at-the-brewery sales pose for the beer category as a whole? What ripples might this shift cause in other parts of the beer market?
- Most of the small brewer organization certifications focus on independence. Why might they choose that over something like product quality?
- What would be your prediction about the growth in the number of breweries and small brewer market share of the next decade? Why? In what ways do you think the market might change? How might consumer demand shift? What would cause those shifts?

9.8 List of References

1. Acitelli T. *The Audacity of Hops: The History of America's Craft Beer Revolution.* Chicago Review Press; 2013.
2. Nilsson I, Reid N, Lehnert M. Geographic patterns of craft breweries at the intraurban scale. The Professional Geographer. 2018;70(1):114–125. doi:10.1080/00330124.2017.1338590
3. Harvard Business Review. The consolidation curve. https://hbr.org/2002/12/the-consolidation-curve
4. Porter M. Clusters and the new economics of competition. https://hbr.org/1998/11/clusters-and-the-new-economics-of-competition
5. Noel J. *Barrel-Aged Stout and Selling out: Goose Island, Anheuser-Busch and How Craft Beer Became Big Business.* Chicago Review Press; 2018.
6. Barajas JM, Boeing G, Wartell J. Neighborhood change, one pint at a time: The impact of local characteristics on craft breweries. In: Chapman NG, Lellock JS, Lippard CD, eds. *Untapped: Exploring the Cultural Dimensions of Craft Beer.* Morgantown, WV: West Virginia University Press; 2017.
7. www.brewersassociation.org, https://www.brewersassociation.org/category/insights/
8. http://www.craftbeer.com/featured-brewery/ncw-albion brewing

Author Biographies

Volker Bornemann, originally from Germany, is founder and principal of Avazyme, a Durham, North Carolina based agriculture, food, and beverage testing laboratory (www.avazyme.com). Volker ensures that Avazyme's scientific knowledge and capabilities are being put to good work for local craft breweries and beverage companies by testing their beers, kombucha, cold brewed coffee etc., as well as troubleshooting their brewery processes. He is an active member of the American Society of Brewing Chemists (ASBC) and the Master Brewers Association of the Americas (MBAA). In his spare time, he is an adjunct professor at North Carolina State University, when he is not kayaking, hiking through the woods with his dogs, or enjoying a cool pint.

Alicia Muñoz Insa studied Agricultural Engineering/Food Engineering at the Technical University of Madrid and did her final thesis at the Technical University of Munich, Weihenstephan, at the department of Brewing Technology. After finalizing her studies, she then started her PhD about sunstruck flavor at the Institute of Brewing and Beverage Technology. Since 2015, Alicia has been working for the Barth-Haas Group as technical manager in England and in Germany.

Christina Schöenberger studied Brewing Technologies at the Technical University of Munich, Weihenstephan. For her doctoral thesis she investigated the sensory importance of non-volatile compounds in bottom fermented beers. She joined the Barth Haas Group in 2005. Since 2011 she is head of the Hops Academy and together with her colleague Georg Drexler she heads the Brewing Solutions department of the Barth Haas Group since 2018. In 2015/16 she was the president of the American Society of Brewing Chemists. She is also a judge of the World Beer Cup, the Beer Sommelier Championship and the European Beer Star Competition.

John Sheppard is a Professor in the Department of Food, Bioprocessing and Nutrition Sciences at North Carolina State University. He received a Ph.D. in Chemical Engineering from McGill University in Montreal, Canada and an MBA from ESCP in Paris, France. Dr. Sheppard currently teaches the science and technology of both biopharmaceuticals and brewing at NCSU. His research is focused on the development and control of fermentation processes, especially applied to yeast. He is also founder and CEO of Lachancea LLC, a company devoted to commercializing novel yeast species. He is a registered PE, a member of the American Society of Brewing Chemists, the Master Brewers Association of the Americas and the NC Craft Brewers Guild. He is also the founder and Director of the NC State University Research Brewery since its inception in 2006.

Bart Watson is Chief Economist at the Brewers Association, a role he has held since 2013. Prior to his position with the BA, he was a lecturer at the University of California, Berkeley, a visiting assistant professor at the University of Iowa, and an associate at The Barthwell Group, a management consulting firm. His academic research focused primarily on the political and economic effects of increasing market consolidation in distribution channels. He holds a B.A. from Stanford University and a Ph.D. from the University of California, Berkeley.

Sebastian Wolfrum is the founder and maltster of Epiphany Craft Malt in Durham, NC, with the mission to provide an integrated regional supply to brewers, distillers, and other sprouted grain ventures. Born and raised in Germany, he earned his craftsmen certification as a brewer & maltster with the Ayinger Brewery outside of Munich from 1997 to 2005. Continuing his brewing career in the U.S., from 2006–2013 Sebastian was Director of Brewing Operations for the second largest craft brewery in North Carolina. As a founding member of the NC Craft Brewers Guild in 2008, he continues to be very involved in the affairs of the small brewers in the region. In 2012, he added a certificate in distilling from IBD, London. In his role as the executive brewmaster, Sebastian joined Capitol Broadcasting Company in 2014 to establish the Rocky Mount Mills Brewery Incubator.

Index

acidity, 159, 160
acquisitions, 191
adjuncts, 16
aeration, 29–31, 41
alcohol content, 155–159
aldehydes, 54
alkalinity, 129–131
ammonia nitrogen, 144
anions, 132
attenuation, 158

barley, 79–85
beer market, 173
beer spoilage, 152–154
beer stone, 28
bicarbonate ion, 129
biochemical oxygen demand, 138
bioreactor, 41
bitterness, 163, 164
boiling, 27–29
bottles, 71
bottom-fermenting, 50
brewery consolidation, 189–192
brewery sanitation, 135–138
brewhouses, 19

brewing fitness, 50
brewing process, 8
business models, 186–189, 193–196

cans, 71, 74–75
carbonation, 61, 68–71
cations, 132
cell counts, 51–53
centrifugation, 68
chitting, 89
clarification, 62, 64–68
cleaning and sanitation, 135
cold crashing, 50, 64
color, 160–162
composition of the dried raw hop
  cone, 106
conditioning, 15
continuous phasing, 44
cooling, 29–30
crabtree-positive, 33
cropping, 50

decoction, 23
demographics, 179–182, 185
deoxynivalenol (DON), 83

diacetyl, 56, 62, 63
distribution laws, 187
double mashing, 24
dry milling, 13

Emil Christian Hansen, 8
enzymes, 16, 80
esters, 54

fed-batch, 43–44
ferment, 5
filtration, 66–67, 76–77
final gravity, 156
fining agents, 65
flavor, 54–57, 150–151, 166–167
foam, 167–171
free-amino nitrogen (FAN), 16, 18,
    47, 143
friabilimeter, 92
fusel alcohols, 55

germ theory, 4
germination energy, 83
germination, 79
greenmalt, 92

hardness, 140
Hayflick limit, 39
heat pasteurization, 75, 76
history of brewing, 3
hop aroma, 114
hop cone, 106–110
hop cultivation, 101–106
hop drying, 104
hop extract, 121
hop oils, 114, 115
hop products, 116–123
hop variety, 112

hops for bittering, 113–116, 131
hops in the brewing process, 27–29,
    111, 113–114
hops, 101

international bitterness units, 163
isomerization, 111–114

Julius Petri, 7
Justus von Liebig, 4

kegs, 71
key performance indicators, 147
kilning, 81

Lachancea thermotolerans, 58
lagering, 62
lambic fermentation, 57
lambic-style, 57
lauter tun, 22
lautering, 22, 26–27
light-induced spoilage, 74
Louis Pasteur, 5

Maillard reactions, 81
malt, 13, 80, 97–98
malting, 79, 86–98
mashing protocol, 19
mashing, 13, 16–26
microbreweries, 174
milling, 13–16
modification, 79
moisture, 82

nucleases, 18

original gravity, 25, 156–159
oxygen, 41, 152–153, 164–166

package option, 71
packaging, 71–75
petri plate, 7
phytase enzymes, 129
phytases, 18
Pierre Berthelot, 6
pitching rate, 49
plumpness, 83
premiumization, 176–179
propagator, 41
protein, 83
pure cultures, 4

real extract, 158
refractometer, 157
Reinheitsgebot, 2
roasting, 94
Robert Koch, 7
Roller mills, 14
run-off rate, 26
rye, 86

Saccharomyces cerevisiae, 33,
    36–37, 57–58
S. pastorianus, 36
Saccharomyces eubayanus, 36–37
secondary fermentation, 61
self-cycling fermentation, 45
soluble proteins, 18
sparging, 23
specific growth rate, 42
spoilage organisms, 152
spread-plate method, 52
steeping, 15, 87
step infusion, 22
surface-active components, 168
sweet wort, 16

taste testing, 150
Theodore Schwann, 5
top-fermenting, 50
total package oxygen, 153
tracking method, 52
trub, 29
turbidity, 162, 163

unit operations, 9

vicinal diketones, 54
vorlaufing, 22

wastewater, 127, 138–146
water analyses, 140–146
water composition, 128–129
water hardness, 129–132, 140–141
water management, 146–148
water quality, 127
water treatment, 132–135,
    138–139
water usage, 127, 130, 146–148
wet milling, 15
wheat, 86
wort composition, 46, 132

yeast cell cycle, 38–39
yeast clades, 35
yeast divergence, 35–36
yeast domestication, 34
yeast flocculation, 64
yeast pitching rate, 49
yeast propagation, 39
yeast reproduction, 37

www.ingramcontent.com/pod-product-compliance
Lightning Source LLC
Chambersburg PA
CBHW050602190326
41458CB00007B/2145

9789811225314